Machine Learning for Edge Computing

Edge AI in Future Computing
Series Editors
Arun Kumar Sangaiah
SCOPE, VIT University, Tamil Nadu
Mamta Mittal
G. B. Pant Government Engineering College, Okhla, New Delhi

Soft Computing Techniques in Engineering, Health, Mathematical and Social Sciences
Pradip Debnath and S. A. Mohiuddine

Machine Learning for Edge Computing: Frameworks, Patterns and Best Practices
Amitoj Singh, Vinay Kukreja, and Taghi Javdani Gandomani

Internet of Things: Frameworks for Enabling and Emerging Technologies
Bharat Bhushan, Sudhir Kumar Sharma, Bhuvan Unhelkar, Muhammad Fazal Ijaz, and Lamia Karim

For more information about this series, please visit: https://www.routledge.com/Edge-AI-in-Future-Computing/book-series/EAIFC

Machine Learning for Edge Computing

Frameworks, Patterns and Best Practices

Edited by

Amitoj Singh, Vinay Kukreja, and Taghi Javdani Gandomani

CRC Press is an imprint of the
Taylor & Francis Group, an **informa** business

First edition published 2023
by CRC Press
6000 Broken Sound Parkway NW, Suite 300, Boca Raton, FL 33487-2742

and by CRC Press
4 Park Square, Milton Park, Abingdon, Oxon, OX14 4RN

CRC Press is an imprint of Taylor & Francis Group, LLC

ISBN: 978-0-367-69432-6 (hbk)
ISBN: 978-0-367-69833-1 (pbk)
ISBN: 978-1-003-14346-8 (ebk)

DOI: 10.1201/9781003143468

Typeset in Times
by KnowledgeWorks Global Ltd.

Contents

Editors

Dr. Amitoj Singh is an Associate Professor in the School of Sciences and Emerging Technologies at Jagat Guru Nanak Dev Punjab State Open University, Punjab, India. He received his Master of Computer Application (MCA) with Distinction from Punjabi University, Patiala and PhD in Computer Science from Punjabi University, Patiala. He has more than 15 years of teaching/research experience; has published more than 30 national/international papers in reputed journals, more than 6 books, and has presented more than 30 papers at national/international conferences. He has supervised 2 PhD students and 15 master's students. He has filed more than 7 Indian patents. His areas of interest include machine learning, software development methodologies, and assistive technologies. He has been granted a Major Research project by IEEE SIGHT to develop language resources for auditory impaired persons. Braille slate developed by him for blind children has won many national and international hackathons and has presented the slate at United Nations headquarters, New York in 2016.

Vinay Kukreja earned his Master's Degree in Computer Science from Punjabi University, Patiala and PhD in Computer Science & Engineering from Chitkara University, Punjab, India. He has been teaching for more than 14 years. He is presently working as an Associate Professor at Chitkara Institute of Engineering and Technology, Chitkara University, Punjab, India. Presently, he is guiding PhD scholars and master's of engineering (ME) scholars. He has also filed patents. He won first prize in the SIH Hackathon (2018) under the flagship of the Ministry of Housing & Urban Affairs, India. His areas of research interest mainly include machine learning, deep learning, agile software development, Natural Language Processing (NLP), data analysis, and structural equation modeling.

Taghi Javdani Gandomani (IEEE senior member) received a PhD in Software Engineering from Universiti Putra Malaysia (UPM), Malaysia, in 2014. He currently serves as an Assistant Professor at Shahrekord University, Shahrekord, Iran. He has about 20 years of work experience in industry and academic institutions. His research interests include software methodologies, software processes, and software improvement.

List of Contributors

Sachin Ahuja
Chitkara University Institute of Engineering & Technology
Chitkara University
Punjab, India

Abhineet Anand
Chitkara University Institute of Engineering & Technology
Chitkara University
Punjab, India

Gagandeep
Department of Computer Science
Punjabi University
Punjab, India

Vinay Gautam
Chitkara University Institute of Engineering & Technology
Chitkara University
Punjab, India

Kalpna Guleria
Chitkara University Institute of Engineering & Technology
Chitkara University
Punjab, India

Deepali Gupta
Chitkara University Institute of Engineering & Technology
Chitkara University
Punjab, India

Kamali Gupta
Chitkara University Institute of Engineering & Technology
Chitkara University
Punjab, India

Gunreet Kaur
Department of Computer Science and Engineering
Thapar Institute of Engineering and Technology
Punjab, India

Prabhjot Kaur
Chitkara University Institute of Engineering & Technology
Chitkara University
Punjab, India

Rajpal Kaur
Department of Commerce and Education
Maharaja Ganga Singh University
Rajasthan, India

Veerpal Kaur
School of Computer Science and Engineering
Lovely Professional University
Punjab, India

Vipul Kaushik
ADDVAL Pvt. Ltd.
Punjab, India

Vinay Kukreja
Chitkara University Institute of Engineering & Technology
Chitkara University
Punjab, India

Umesh Kumar Lilhore
Chitkara University Institute of Engineering & Technology
Chitkara University
Punjab, India

Anand Muni Mishra
Chitkara University Institute of Engineering & Technology
Chitkara University
Punjab, India

Aditi Moudgil
Chitkara University Institute of Engineering & Technology
Chitkara University
Punjab, India

Huma Naz
Chitkara University Institute of Engineering & Technology
Chitkara University
Punjab, India

Madhavi Popli
Department of Computer Science
Punjabi University
Punjab, India

Ram Kumar Ketti Ramachandran
Chitkara University Institute of Engineering & Technology
Chitkara University
Punjab, India

Manisha Rani
Research Scholar
Department of Computer Science
Punjabi University
Punjab, India

Monika Sethi
Chitkara University Institute of Engineering & Technology
Chitkara University
Punjab, India

Neha Sharma
Chitkara University Institute of Engineering & Technology
Chitkara University
Punjab, India

Rishabh Sharma
Chitkara University Institute of Engineering & Technology
Chitkara University
Punjab, India

Saravjeet Singh
Chitkara University Institute of Engineering and Technology
Chitkara University
Punjab, India

Naresh Kumar Trivedi
Chitkara University Institute of Engineering & Technology
Chitkara University
Punjab, India

Shivani Wadhwa
Department of Computer Science
Punjabi University
Punjab, India

1 Fog Computing and Its Security Challenges

Kamali Gupta, Deepali Gupta, and Vinay Kukreja
Chitkara University Institute of Engineering & Technology, Chitkara University
Punjab, India

Vipul Kaushik
ADDVAL Pvt. Ltd
Punjab, India

CONTENTS

DOI: 10.1201/9781003143468-1

1.1 INTRODUCTION TO FOG COMPUTING

1.1.1 Introduction

The word fog originates from the term "clouds at the edge," [1] which is symbolized as services offered to users near to the ground. Fog [2] is predominately envisaged to services of cloud through a unified Application Programming Interface.

The paradigm of fog [3] is that it is a distributed and vulnerable environment that is well equipped with provisions of storage facilities, computational mechanisms, and networking methodologies and facilitators.

1.1.2 Architecture

The architecture of fog is hierarchical and bi-directional in its service offerings. Figure 1.1 presents an architecture of the fog model that briefly discusses that data from devices and sensors gets generated in large quantities and is further routed to the fog node, where it is reduced to eradicate unnecessary data. The nodes then send this data to cloud for the purpose of decision-making. Therefore, the fog layer acts as a filter that removes unwanted patterns specific to the application/request in hand. Thus, the three-layer architecture is a robust process-management system that controls and manages data for its efficacious use [4, 5]. Figure 1.2 illustrates the composition/components of these three layers, wherein the sensors are deployed at edge of the network. Cloudlets are at the Fog layer and Cloud is at the top-most level.

The next section embarks on the characteristics of fog model.

1.1.3 Characteristics

Fog as an emerging paradigm is able to extend a number of advantages in comparison to cloud, which is a decentralized network. The parametric representation of characteristics of cloud and fog presented in Table 1.1 is a clear depiction of its popularity in a number of real-life applications that are discussed subsequently.

1.1.4 Applications

Although the cloud model has captured the market in its entirety, it still suffers from increased latency due to massive traffic. This has led to fog playing a major role in

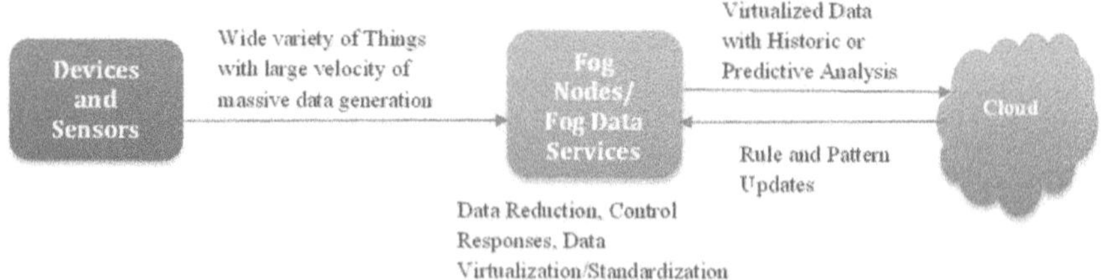

FIGURE 1.1 Wide variety of things with large velocity of massive data generation.

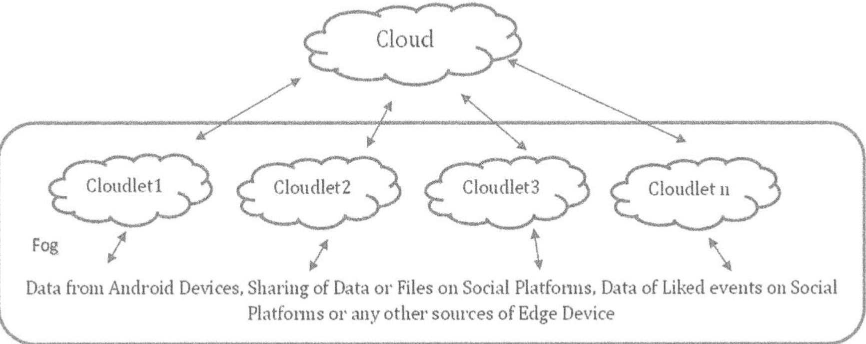

FIGURE 1.2 Fog computing: A depiction of three layers.

catering to the requirements of *fog applications in the real world* [7], such as smart cities, smart grids, smart homes, healthcare, oil and gas, connected vehicles, agriculture, energy management, retail transportation, wireless sensor networks, software defined networks, Internet of Things (IoT) and cyber-crime physical systems, mobile computing systems, and many more.

This chapter has been organized considering the fog as an emerging technology and the benefits it proffers in the paradigm of computing. Efforts have been made to expound on the existing fog offerings pertaining to its definition, characteristics, advantages, workings, security challenges, and security solutions. A summary based on a huge, dedicated survey has been drawn as a conclusion, and observations and future directions are further suggested. The glimpse of the *organizational structure* of this chapter is illustrated in Figure 1.3.

TABLE 1.1
Characteristic Parameters of Fog Computing [6]

S.No.	Parameter/Characteristic	Cloud Computing	Fog Computing
1.	Latency	High	Low
2.	Delay jitter	High	Very low
3.	Location of service	Within the internet	At the edge of local network
4.	Distance between client and the server	Multiple hops	One hop
5.	Security	Undefined	Can be defined
6.	Attack on data en route	High probability	Very low probability
7.	Geo-distribution	Centralized	Distributed
8.	Support for mobility	Limited	Supported
9.	Real-time interactions	Supported	Supported

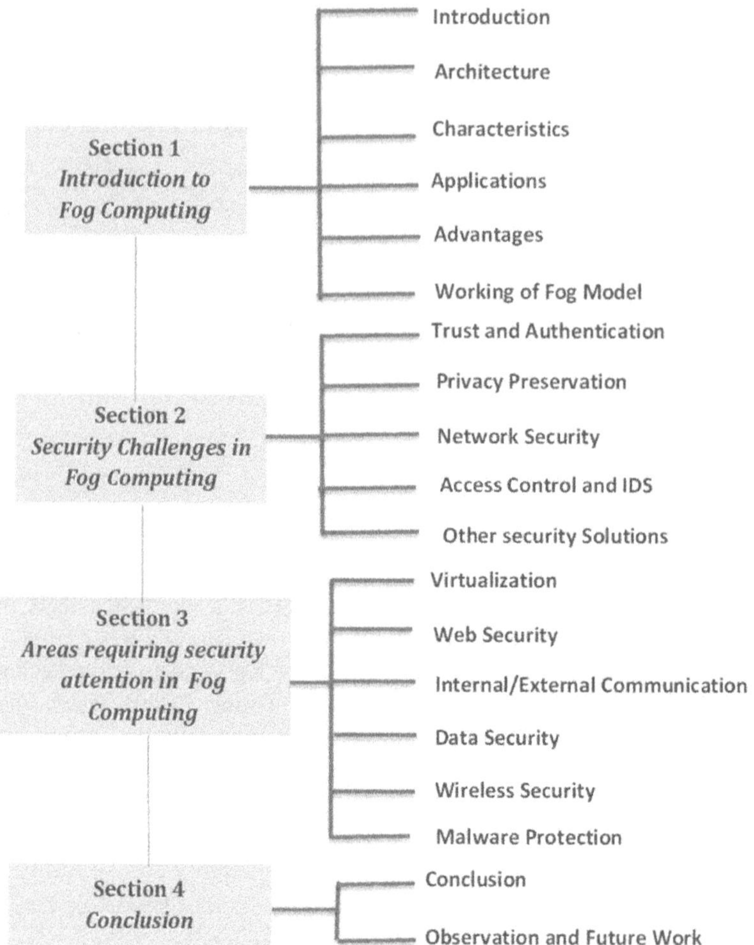

FIGURE 1.3 Organization of this chapter.

1.1.5 Advantages of Fog Computing

The advantages of the fog model can be understood by analyzing its applications such as smart architectures, which include buildings, grids, Software-Defined Networking (SDN), and cities that completely exploit all its services for better service offerings [8].

The prime advantages of the fog model include a better response time with the system, reduced/filtered traffic directed to the cloud platform, a reduced usage of bandwidth, mobility support, greater scalability and security, portability, the handling of heterogeneity from all dimensions, localized control management of massive data, and so on.

Thus, looking at the advantages offered by the fog paradigm in the world of computing, several technology enablers are shifting to its methodologies. A glimpse of such vendors appears in Figure 1.4 [9].

The next section expounds on the working methodology of fog computing.

Technology Enabler for Fog Computing	Cloud Service Provider such as vmware or amazon.com
	IoT Device Manufacturers like Google, Microsoft, Apple
	Computer Chip suppliers like Qualcomm, Intel
	End User Experience Provider like GE, Toyota, BMW
	Network Operator like verizon, comcast
	Network Equipment Vendor like Nokia, CISCO, Ericsson
	System Integrator like IBM, HP
	Edge Device Manufacturer like Linksys

FIGURE 1.4 Technology enablers for fog computing.

1.1.6 Workings of the Fog Model

The fog acts as a library for implementing cloud services and is written in Ruby language. The steps that embark on its working methodology are presented below:

- At the outset, the cloud center is set up to fix up resource specifications such as the internet, storage, computing, networking, and so on. The cloud server is switched on in the cloud environment.
- Subsequently, the fog layer is configured and resides at the edge by switching on the fog server using the fog file as it has been stated in working methodologies.
- The user submits his request compositions for service acquisition.
- The compute model runs with respect to the technology enabler for which the request has been generated using technology employed in the back end of the fog model.

The above steps outline the general workings of the fog model and are acceptable to modification as per application-specific requirements. Though the fog model caters to eradicate the effects of latency in the network, it still gets exposed to security threats. Therefore, challenges related to fog security are discussed in the next section, along with their anticipated solutions.

1.2 SECURITY CHALLENGES IN THE FOG MODEL

The data that is being routed to different machines while traversing from fog to cloud faces huge security challenges. It is imperative that these security challenges are handled and security solutions be customized as per the threats encountered. Figure 1.5 illustrates some of the security challenges being faced by the fog model.

Each challenge is described in and its categorization is presented in the following sections.

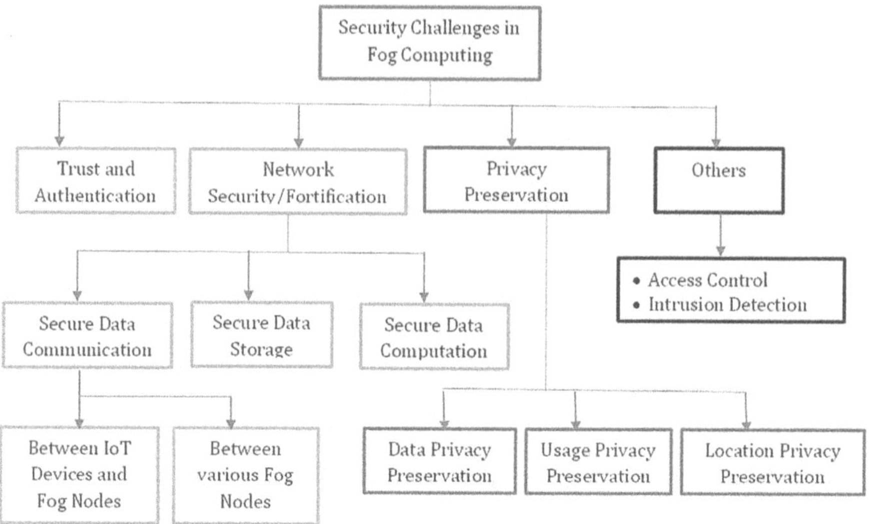

FIGURE 1.5 Security challenges in fog computing.

1.2.1 Trust and Authentication

1.2.1.1 Trust

The networking capabilities [10] offered over fog are expected to be reliable and secure. This can be ensured if the participating devices maintain a degree of trust with one another. Authentication is a secure way to validate the identity of communicating devices by establishing a relation between the fog network nodes and IoT devices. Though the authentication parameter is beneficial in permitting authentic users, the network is still susceptible to malfunction due to malicious attacks. A possible solution to the security issue can be offered by the trust as it can foster relations on the basis of previous interactions. It comprehends solutions by validating if the fog node offering service to the IoT device is genuine. A fog node is the most crucial component as it is responsible for ensuring privacy in the network. Further, the IoT devices aiming to communicate with the intended fog nodes confirm the security of the data and request flow prior to their communication. The research works presented in [10–12] address the issues pertaining to trust in network communications.

The issues encountered in the implementation of trust on the fog model can be understood more clearly by looking at some questions and their relevant answers:

- How do you recognize the attributes and definition of trust with respect to the fog environment?
  Solution: Feedback of the previous runs.

It is imperative to recognize the attributes and definition of trust with respect to the fog model, though a number of solutions have already been comprehended in the cloud environment. But since they have been derived to provide viable solutions at centralized levels, so to cope up with mobility related issues, techniques need to be proposed that lend feasible solutions in the fog model. One such research has been carried out in this direction and is available in [13]. The focus of this research has been directed toward evaluating the main attributes that can comprehend the definition and mechanism used to measure trust in a fog model. The researcher sets up an experimental test bed comprising 18 service providers to measure their trust value in terms of their previous service offerings. The research first evaluates the performance of service providers in terms of trust before assigning them any request appearing in the fog environment. The researcher has presented a hybrid mode that can cater to the issue of measuring trust to evaluate the performance of cloud providers.

The model implementation has been realized by first collecting and aggregating the feedback from the service providers of previous runs. The aggregated value is termed as trust value and is oriented toward choosing the best service provider among a set of available ones for the submitted user request. The research work shows a complete mapping of the request to service fulfillment as a series of steps incepted from the setup of the cloud and fog center to the evaluation of service providers by the evaluator module. The research work contributes to the maintenance of service level agreement (SLA) between customers and service providers. Similarly, the solution needs to be derived for the networks that are being built arbitrarily. This issue is discussed in the question below:

- How do you ensure trust in services offered by vendors capable of building fog networks capriciously?
  Solution: Reputation based trust model.

Besides, as communicated in [14], the fog model and the cloud model use different approaches in the ownership of data centers. The fog model offers flexibility in deployment choices to vendors capable of offering a particular service type while the cloud model is implemented as a usual mechanism a data center is owned by a service provider. The fog model categorizes the vendors building their own fog networks as follows:

- **Internet Service Providers:** They are also termed as wireless carriers as they have been actively involved in controlling cellular base stations or home gateways and can build their own networks with the existing infrastructural resources.
- **Cloud Service Providers:** They comprise a group of people concerned with expanding their IT services to the edge of networks by establishing their own fog networks.
- **End Users:** This constitutes a group of people who want to lease their spare resources by converting their local private networks into typical fog models in order to decrease the cost of ownership.

The flexibility of making fog networks capriciously complicates the implementation of trust in fog. In order to inculcate trust level into communications that are peer-to-peer (P2P) centric, involve e-commerce, and are based on user reviews and meant for social networking, robust reputation systems are efficacious. In this direction, a robust study has been carried out in [15], wherein the researcher has proposed a robust reputation system for implementing the task of resource selection in P2P networks. A distributed polling algorithm has been developed that measures the reliability of a resource prior to its usage on internet. Efforts have been put forth to address issues related to distinct and unique identity, differentiation, and the handling of intended and accidental access and redemption and castigation of reputation. The author has tried to implement a self-regulating P2P system that can isolate any node depicting any illegal behavior in the system.

1.2.1.2 Authentication

The authentication process is the first step of security implementation in a fog environment. It keeps check on the entities entering the cloud/fog model to extract or offer services in the form of users and providers, respectively. Besides implementing security in cloud, the authentication process also addresses the issue of resource acquisition and ensures uninterrupted service to the user. The research work presented in [10] addresses the following two issues more clearly:

1. Since massive traffic lands to fog to extract the services limited in quantity, it is imperative to restrict the entry of unauthorized nodes aspiring to acquire the resources from entering to the fog.
2. The dynamism property of cloud helping the user acquire resources as needed offers flexibility in frequently leaving and joining the environment, which poses interrupted services to the users who are registered in fog. Some measures need to be adopted that can enforce the complexity in the registration and re-authentication process for users logging in frequently within a framed period of time.

   Solutions: Traditional techniques of authentication involving Public Key Infrastructure (PKI) and certificates experience scalability and resource constraints. In today's scenario, with increasing traffic demands to establish communication among a number of people, achieving a desired level of security and privacy raises a major concern as it becomes difficult to assess the intentions of users entering into cloud. Several studies have been conducted in this context and are discussed as follows:

- **Public Key Infrastructure:** A PKI-based authentication protocol has been proposed in [16] that utilizes the technique of multicast authentication for enabling secure communications.
- **Multicast Authentication Protocol:** In order to cater to security issues originating in multicast communications, a study has been explicated in [17]. The research paper elucidates a new multicast authentication scheme

for real-time applications using an advanced encryption standard algorithm by assigning a new index number to an entrusted member whenever a block of packet is being sent in a multicast group.

- **Intermediary Certifying Authority:** Another solution to put a constraint on the service requests arriving from malicious and compromised nodes by restricting unauthorized nodes to become a part of the fog network can be assured by involving an intermediary certifying authority that can authenticate the requesting devices aspiring to obtain services of storage and processes.
- **Fine-Grained Access Control:** Authenticity can also be achieved by employing fine-grained access control mechanisms, as illustrated in [18]. Cryptographic techniques are employed here to assign an individual access control scheme to each data item owner. These access control policies are kept secret as the attributes specific to control policies defined by the data owners responsible for associating mapping between the attributes and their keys are not revealed to the holders of the credential keys. Similarly, biometric authentication has been emerging as a good solution in mobile and cloud computing. Research is being done to implement fingerprint, face, touch-based, or keystroke-based authentications in the area of fog computing to enhance data privacy.

Some more models explored by research work carried out in [10] are expounded in Table 1.2.

1.2.2 Privacy or Confidentiality Preservation

When the services are accessed over cloud by massive traffic, they generate a threat to access and leak the information overshadowing the security models implemented in cloud. The data residing and exchanged between fog clients is more prone to unauthorized access as the fog nodes are closer in vicinity to end users compared to remote cloud nodes [14].

To implement high levels of security in fog, privacy needs to be addressed at various levels. There is a dire need to look into the following questions to ensure privacy:

- How can data privacy be ensured in a typical fog model?
- How can a fog service usage pattern be composed for a fog client in order to ensure maximum privacy?
- How can location privacy be ensured at the time a fog node offloads tasks to its nearest neighbor while assuming other nodes to be distant?

The above questions pose needs to fog model implementers to derive feasible solutions that maintain data, usage, and location privacy. A brief layout of the content covered under this section is presented in Figure 1.6. The techniques are discussed in detail later in this chapter.

TABLE 1.2
Existing Authentication Schemes for Fog Computing

System Model	Authentication and Privacy Models	Countermeasure	Performances (+) and Limitations (-)
Fog computing with face identification and resolution application	• Confidentiality • Integrity • Availability	• Authentication and session key agreement • Advanced Encryption Standard (AES) symmetric key encryption mechanism based on session key • Secure Hash Algorithms (SHA-I) algorithm	(+) Response time for different face databases (+) Can detect Man-in-the-Middle (MITM) attack and identify forgery (-) Increased computation and communication overhead a bit
Fog computing-enhanced IoT	• Privacy of individual IoT device data • Integrity	• Chinese reminder theorem • Homomorphic Paillier encryption • One-way hash chain	(+) Computational cost and communication overhead compared to the aggregation (+) Can resist against the false data injection attack (+) Fault tolerance (-) Traceability is not considered
Vehicular crowd sensing using fog computing	• Confidentiality • Mutual authenticity • Integrity • Privacy • Anonymity	Certificate less aggregate	(+) Computational cost and communication overhead (+) Key escrow resilience (+) Provide anonymity compared to the scheme (-) Location privacy is not considered
Fog storage architecture with the three system entities, including the cloud, fog, and end user	• Data privacy • Forward secrecy	• Printing-based cryptographic • Merkle tree • User-level key management and update mechanisms	(+) Secure user-level key management (+) Efficient in terms of computation, communication, and storage as compared to the scheme (-) Adversary's model is limited

(Continued)

TABLE 1.2 *(Continued)*
Existing Authentication Schemes for Fog Computing

System Model	Authentication and Privacy Models	Countermeasure	Performances (+) and Limitations (-)
Vehicular ad hoc network (VANET) using fog computing	• Identity privacy • Location privacy • Authenticity	• Location based encryption (LBE) scheme • Cryptographic puzzle • SHA-I algorithm	(+) Average delay to solve a puzzle (+) Average delay to verify the proofs (+) Defending denial-of-service attacks (-) Anonymity is not considered compared to the scheme (-) Adversary's model is limited
Location-based fog computing	• Location privacy	• Position based cryptography (e.g., position-based key exchange) • Position-based multi-party computation, position-based public key infrastructure	(+) Without using additional computation overhead (-) Integrity is not considered
Fog data centers consist of four renewable fog nodes	• Differential privacy	• Query function • Laplacian mechanism	(+) Efficient in terms of execution efficiency, private preserving, quality, data utility, and energy consumption (+) Resist node and edge recognition attack (-) Adversary's model is limited

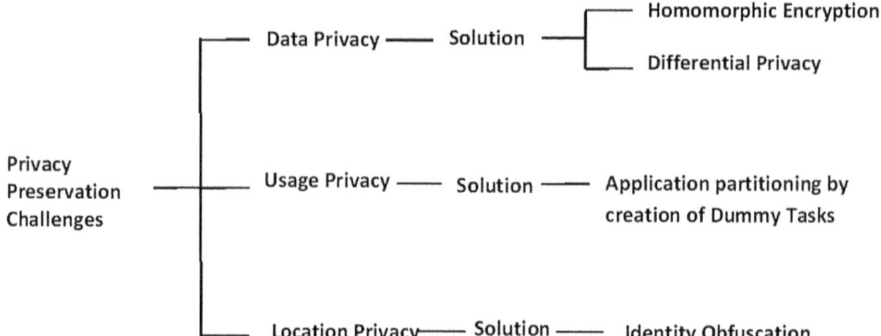

FIGURE 1.6 Privacy preservation challenges and solutions.

1.2.2.1 Data Privacy Preservation

Though the fog networks have a number of privacy-preserving algorithms residing between the cloud and the fog model that have restricted access to resources lying at the edge, these algorithms can still collect data sensitive in nature produced by sensors and other end devices. The solution to this stated problem could be to adopt techniques of *homomorphic encryption* and *differential privacy.*

Homomorphic encryption, a popular encryption technique, allows computations to be performed on ciphertexts, creating an illusion that the data is still in its original form [19, 20]. It lets the user perform complex mathematical operations on encrypted data, which subsequently generates an encrypted result that, when further decrypted, is similar to the result of operations when performed on plain text. It carries all the operations efficaciously without compromising the encryption applied. The term has derived its meaning from the Greek word "same structure." It is typically being used in systems such as collision-resistant hash functions, secure voting systems, private-set interactions, predictive analysis in healthcare, and many more.

Differential privacy refers to non-revelation of privacy by any single entry belonging to any dataset when performing statistical queries [21]. It is a statistical technique that optimizes the performance of query evaluation in statistical databases. The technique was developed by cryptographers and can be explained by this example: Differential policy when applied on two identical databases, with the former containing some information pertaining to the user not contained in the later one, ensures that any statistical query fired on both will produce nearly same result irrespective of the database from which the query has been fired.

1.2.2.2 Usage Privacy Preservation

This privacy can reveal information pertaining to user service usage patterns. Although usage information is needed to match the SLA requirements, it can be harmful in disclosing information sensitive to other user aspects. For example, with a smart grid, the readings in a smart meter not only tell the required electrical usage information but also can breach user privacy by disclosing information such as the time a user is not at home, and so on. In order to take care of such issues, solutions can be provided using a trusted third party that can employ a mechanism to ensure only information relevant to the applied application is revealed. This can be achieved by creating *dummy tasks* by the fog client and offloading them to multiple assorted fog nodes, thereby hiding the real clubbed tasks. Though it offers a good solution because each fog node is concerned only with its stated task, it can waste resources and increase the financial burden on the fog client. Therefore, a smart mechanism to partition the applications can be defined to ensure resource offloading without any disclosure of sensitive information. The technique is recommended for future research.

1.2.2.3 Location Privacy Preservation

The resources available over the fog/cloud model are limited in number, which raises the provider's need for context-aware sharing among devices that are geographically closer in order to increase resource utilization. Such sharing can be prone to a Man-in-the-Middle (MITM) attack that can access and disclose massive amount of data sensitive in terms of location, identity pertaining to fog nodes, and so on, as per [22, 23]. The location privacy can be preserved by *identity obfuscation*. It indicates that even though the communicating fog nodes are neighbors, identity verification is still needed in order to maintain and establish their communication. Further, practices such as offloading onto some other fog client on the basis of not only location, but also latency, load balance, and reputation can create a feasible solution, as the node desiring to offload will not have a precise idea about the location of the desired fog node, restricting the communication in intersection of their locations.

Though addressing solutions to these three stated areas can provide some good privacy implementations, there are other security issues related to privacy as proposed by the researchers in [10]. The issues along with various questions to be addressed are presented in Figure 1.7. The colored part in the figure resembles the attributes pertaining to privacy challenges in security, and the question to be resolved is stated under the attribute.

1.2.3 Network Security

The notion [14] of fog computing has realized its implementation with the emergence of wireless technology. This has raised major concerns for implementers with wireless network security as it may be affected due to attacks such as jamming attacks, sniffer attacks, and so on. The virtuous practice of network management is isolating regular data traffic from network management traffic [24]. As the fog nodes reside at the edge of the network, it muddles up the network management process due to an increase in the cost of managing huge scale servers distributed in the network in its entirety with reduced access for maintenance. The problem can be eradicated by employing *SDN* technique. SDN is a Software-Defined Networking technique capable enough to manage an entire network by isolating network control functions from forwarding functions of a network [7], that is, SDN is a networking paradigm that segregates the task of data control from the data communication layers [25]. The control functions can include the task of routing, application execution and related work, policy definitions, and so on. The decision regarding routing traffic is solely deputed to a virtual network control plane while SDN abstracts the

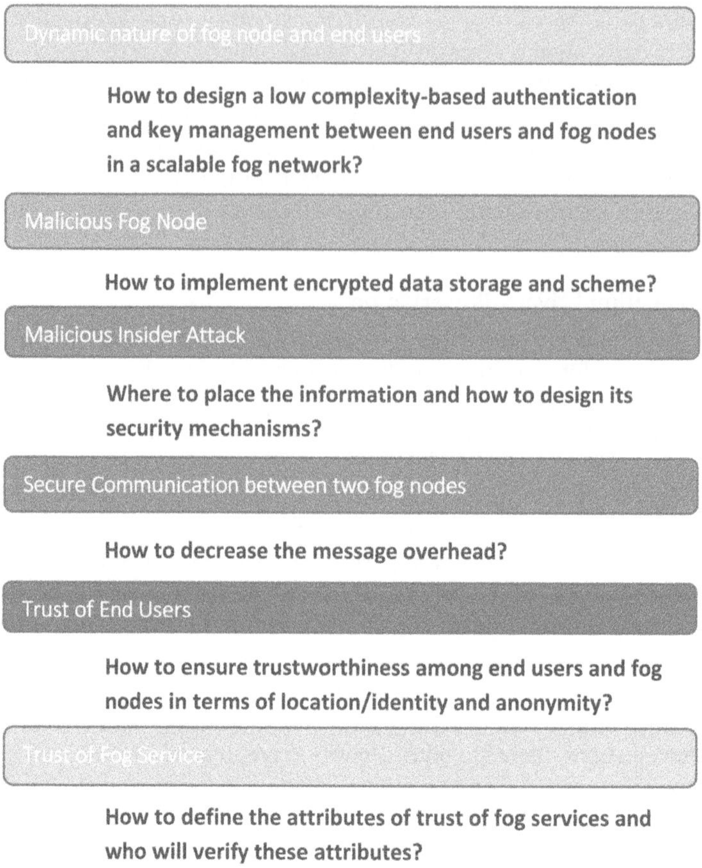

Dynamic nature of fog node and end users

How to design a low complexity-based authentication and key management between end users and fog nodes in a scalable fog network?

Malicious Fog Node

How to implement encrypted data storage and scheme?

Malicious Insider Attack

Where to place the information and how to design its security mechanisms?

Secure Communication between two fog nodes

How to decrease the message overhead?

Trust of End Users

How to ensure trustworthiness among end users and fog nodes in terms of location/identity and anonymity?

Trust of Fog Service

How to define the attributes of trust of fog services and who will verify these attributes?

FIGURE 1.7 Security and privacy issues in fog computing.

physical networking resources. Thus, SDN plays a major role in network management, increases scalability, and reduces cost. SDN in association with fog computing [26] is providing solutions in implementations of vehicular networks, resolving collisions and high-packet loss rates, and intermittent connectivity by augmenting vehicle-to-infrastructure and vehicle-to-vehicle communications.

The traits of network security can be understood and justified while looking at and answering the following questions:

- Is the data stored in the fog environment secure?
- Are the computations performed on this stored data secure?
- How is data that is in transit of its flow secured?

Therefore, dealing with the issue of network security necessitates to deal with the underlying issues of secure data storage, secure data computation, and secure data communication. These issues are briefly discussed here.

1.2.3.1 Secure Data Storage

In fog networking, the outsourcing of user data is done as the user's control over fog data gets delegated to fog node. This introduces the issue of ensuring data integrity as the outsourced data may get tailored accidentally or may get lost. Additionally, the data could be misused by unapproved parties for a third person. To address these problems, *auditable data storage services* are suggested. The providers of services can opt for *homomorphic and searchable encryption* in combination to inculcate confidentiality, integrity, and verifiability for data residing on untrusted servers. Further, a *privacy-preserving public auditing mechanism* can be employed for data that relies on a Third- Party Auditor (TPA) for using the techniques of a homomorphic authenticator or random mask method, as suggested in [27].

1.2.3.2 Secure Data Computation

The aim of fog computing [7] is to minimize the traffic deviating to cloud for the acquisition of services in the form of resources for its computational needs. This has posed concerns dealing with node heterogeneity in computation, automatic troubleshooting, automatic deployment, service discovery, intelligent configurations, and so on. Dealing with such outsourcing of computation has led to the discovery of possible solutions such as *verifiable computing* and *data search*. Verifiable computing aims at instilling user confidence in the computations being delegated to fog nodes as the user can verify the correctness of the computations. It allows computations to be offloaded to untrusted servers despite maintaining verifiable results. The servers other than the offloaded one evaluate the process and revert with a proof that the result is correct. In this context, the research work presented in [28] complements a significant verifiable computing protocol capable of returning a computationally-sound proof that can be verified by the client with no additional cost. Similarly, the work presented in [29] provides an absolute demonstration of verifiable computing by producing a model, named "Pinocchio," which allows the client to create a public evaluation key that evaluates general computations performed by a server taking into consideration cryptographic assumptions. To ensure data privacy, the prerequisite for outsourcing sensitive data is an encryption that checks for data utilization and deployment services. The champion of all these techniques is the *keyword search* technique that works on patterns in the form of keywords searching encrypted files. The research work conducted in [30] elucidates schemes that perform searches on encoded data for provable secrecy by query isolation, provision for hidden query, controlled searching, and so on.

1.2.3.3 Secure Data Communication

Though security implementations [10] are required in a network while data is in phase of storage and computation, data is prone to exposure while in the communication phase. IoT devices are required to have some security mechanisms to provide security as data cannot be offloaded to fog devices lying in its vicinity. Communication is established in a fog network for the purpose of offloading either a storage or a computation request. The fog nodes interact with each other to discuss issues pertaining to resource management or network management. Thus, the interactions in a typical fog model are either established between IoT devices and fog nodes or between various fog nodes. In order to ensure secure data communication, the communication enablers are required to look at these three perspectives:

- How can the communication between fog nodes and IoT devices be secured?
- How can the communication between various fog nodes be secured?
- How can the overhead of beckon and other such type of messages be reduced within resource constrained networks?

An IoT device aspiring for a service is sometimes not aware of the existence of any fog network. This may create an issue of unsecure communication as symmetric cryptographic techniques may not be able to offer good protection. Asymmetric key cryptography and implementing PKI techniques to foster security in the network have their own challenges. Additionally, network users also suffer from the bogus messages circulated in the network by attackers. Solutions to such an issue have been provided in the research work presented in [31]. An intermittent and flexible security mechanism has been devised that can cater to the problem of unreliable network connections by establishing security configurations particular to application requirements. Further, techniques of *homomorphic encryption* can be utilized that use pseudonyms to guarantee data secrecy by protecting the identity of devices lying at the edge of the network [32]. To maintain the integrity of data, some kind of *masking algorithm or light-weight encryption* technique can be used. Such techniques make use of lightweight block ciphers, stream ciphers, hash functions, and similar methods to measure security without compromising limited resources.

A summary of various perspectives of secure data communication, along with their solution technique, is presented in Table 1.3. In addition, there are some other challenges that are presented in Section 1.2.4.

1.2.4 Others

The security challenges that also affect the performance of a fog model are presented as follows:

a. **Access Control:** The outsourcing of data and services in cloud requires a pre-check on access control and can be implemented using cryptographic techniques. Methodologies such as fine-grained access control and policy-based

TABLE 1.3
Perspectives and Techniques for Secure Data Communication

Perspective for Secure Data Communication	Technique That Can Foster a Good Solution
Secure data storage	Homomorphic encryption, privacy preserving public auditing
Secure data computation	Verifiable computing, data search
Secure data communication	Homomorphic encryption, light-weight encryption

access control can be used for secure operations between heterogeneous fog clients and IoT devices. Several approaches can be used to implement better access control in a fog environment and are presented as follows [33]:

- *Behavior Profiling*: It deals with protecting the data from unauthorized access. It involves behavior profiling a user via decoying information technology methodologies to compare the behavior of an unauthorized user to an authorized one.
- *Attribute Encryption*: To ensure the security of data, advanced encryption techniques such as the advanced encryption standard algorithm, the Rivest Shamir Adleman algorithm, and the ciphertext attribute-based encryption algorithm can be used. Such attribute-based approaches guarantee confidentiality and fine-grained access control.
- *Certifying Authority and Policy-based Access Control*: For enhanced security in fog-based networks, certificate-based revocation and information distribution need to be supported by fog nodes. It involves entities such as a back-end cloud, certificate authority, IoT devices, and fog nodes. Extensible access control mark-up language can be used to formalize network policy specifications for supporting secure collaborations and interoperability among the resources heterogeneous in nature.

b. **Intrusion Detection:** The task of these systems [10] is to mitigate attacks like Denial of Service (DoS) attacks, scanning attacks, insider attacks, MITM attacks, and so on. These detection methods are applicable to systems such as cloud, smart grid, supervisory control and data acquisition (SCADA), and so on. Fog computing, as discussed earlier, follows a three-tier architecture. It is imperative to apply these detection techniques to all the three layers of a fog model for analyzing and monitoring traffic and the attributes of end devices, fog nodes, and cloud servers. In such multi-layered architectures, implementing security in one layer does not verify the propagation of malware from a vulnerable node to others. There is a dire need to deploy perimeter Intrusion Detection Systems (IDSs) to coordinate assorted detection components for improvised security implementations. These can better assist in dealing with challenges such as alarm parallelization, real-time notifications, correct responses, and false alarm control.

In cloud computing, IDS [14] plays a vital role in detecting attacks such as flooding attacks, insider attacks, port scanning, attacks on hypervisors, and so on. In fog computing, IDSs are used to check log files and user login information, control policies, and detect DoS attacks. Though IDSs are well-implemented in fog and cloud architectures, a lot of efforts still need to be made with mobility, massive traffic, geo-distribution, latency minimization, and so on.

1.2.4.1 Other Security Solutions

Section 1.2.4 discussed in detail possible threats to data when it is in various phases residing in a fog along with some possible solutions. In addition, solutions that can be extended to resolve any challenge along with their benefit are presented in Table 1.4.

TABLE 1.4
Efficacious Solutions to Security Challenges [34]

Solution Category	Security Challenge Resolved	Benefit Offered
Data encryption	• Malicious insiders • Data loss • Spyware/malicious processes • Insufficient due diligence • Data breach	It will protect data from unauthorized access even though the data is breached either at the time of computation, rest, or transit.
Network monitoring	• Advanced persistent threats • Denial of service attack • Access control issues • Abuse and nefarious use of resources • Malicious insiders • Data breaches • Insufficient due diligence • Attack detection	• Logging of nasty events for later analysis • Indicate behavioral attributes of system with respect to security parameters • Block apprehensive incoming/outgoing traffic • Instant notification about ongoing attack
Malware protection	• Insecure API • Account hijacking • Data corruption/damage risks • Service and application vulnerabilities • Performance degradation • Shared technology issues	Provisions for scanning and exclusion of malicious applications in real-time
Wireless security	• Access control issues • Advance persistent threats • Data breach • Illegal bandwidth consumption • Eavesdropping attacks	Allows scalability in connection of fog devices as per their availability in a better secure manner
Secured multi-tenancy	• Account hijacking • Access control issues • Insecure APIs • Abuse and nefarious use of resources • Malicious insiders • Segregation issues • Data breaches	• Efficacious utilization of fog resources • Avoidance of hopping attacks • Secure data associations between authorized users
Backup & recovery	• Data unavailability issues • Data loss • Insufficient due diligence • Data integrity issues • Malware infection	The availability and integrity of data is preserved even at the time of disasters
Securing vehicular networks	• Access control issues • Advance persistent threats • Account/session hijacking • User identity protection • Denial of service attacks	It preserves the identity of the user and location to increase the road safety by preserving data in communication

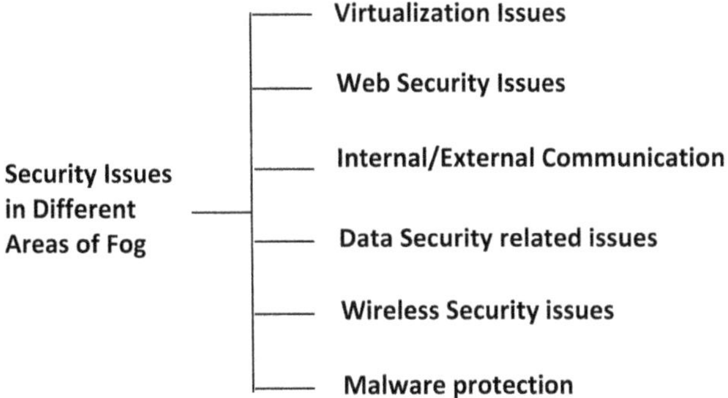

FIGURE 1.8 Areas desiring security attention in fog.

1.3 AREAS REQUIRING SECURITY ATTENTION IN FOG

In addition to the security challenges discussed previously, the areas that require security to be implemented are presented in Figure 1.8, and illustrated in the survey carried out in [34, 35].

The entire domain of the computing paradigm depends on robust software and the processes adopted for its development [36]. Additionally, when developing any agile project that is secured in its operation, structural equation modeling is used that can help facilitate a timely completion of projects. So, from the point of view of how submitted requests in the computing environment are better served besides being secured, agility is a major component.

Areas desiring security attention fog are presented in Table 1.5.

Table 1.5 presents various areas in a typical fog paradigm where implications of security necessitate having better mechanisms that can cope with the possible threats. The table also explicates possible solutions in this direction along with the impact of the threat if any lapse occurs in security implementation. Section 1.4 presents the conclusion on the conducted survey.

1.4 CONCLUSION

This chapter analyzes fog paradigm security flaws along with their existing solutions. It was observed that research practices adopted by various researchers focus on improving fog in terms of its functionality, and a few research works focus on fog security parameter. Consequently, the whole research work explicates various security challenges, possible solutions to them, IoT areas requiring security attention in fog, and some of the generalized solution techniques with their benefits. The research presented in this chapter helps readers understand fog security measures in depth while envisioning the architectures of novel fog models.

TABLE 1.5
Areas Desiring Security in Fog Computing [34]

Fog Security Area	Possible Threats	Possible Solutions	Impact
Virtualization	• Hypervisor attacks • Side channel attacks • Privilege escalation • Service abuse • Inefficient resource policies • Privilege escalation attacks • Virtual machine (VM)-based attacks	• Multi-factor Authentication • IDS • User data isolation • Role-based access control model • Attribute or identity-based encryption • User-based permissions model • Process isolation	Each computation is taking place in a virtualized environment, so any lapse in security implementation may lead to adverse effect on every fog service.
Wireless security	• Message replay attacks • Message distortion issues • Data loss/breach • Illegal resource consumption • Sniffing attacks • Active impersonation	• Encrypted communication • Authentication • Secure routing • Wireless security protocols • Key management service • Private network	Any lapse in security implementation can compromise consistency, availability, accuracy, trustworthiness, and privacy.
Internal/external communication	• MITM attack • Poor access control • Inefficient rules/policies • Insecure APIs and services • Session/account hijacking • Single point of failure • Application vulnerabilities	• Encrypted communication • Partial encryption • Mutual/multi-factor authentication • Isolating compromised nodes • Transport layer security • Limiting number of connections	Any lapse in its implementation could lead to eavesdropping by an attacker to access constrained resources.
Web security	• Cross-site scripting/ request forgery • SQL injection • Insecure direct object references • Session/account hijacking • Drive-by attacks • Malicious redirections	• Secure code • Regular software updates • Find and patch vulnerabilities • Periodic auditing • Intrusion prevention system • Firewall/anti-virus protection	Any lapse in implementing security can enable the attacker to get into the system and install malicious application.
Malware protection	• Virus • Worms • Trojans • Spyware • Performance reduction	• Anti-malware programs • IDS • Rigorous data backups • System restore points • Patching vulnerabilities	Any lapse could result in damage of data permanently and degradation of system performance.

(Continued)

TABLE 1.5 (*Continued*)
Areas Desiring Security in Fog Computing [34]

Fog Security Area	Possible Threats	Possible Solutions	Impact
Data security	• Data altering and erasing attacks • Data replication and sharing • Data ownership issues • Illegal data access • Low attack tolerance • Multi-tenancy issues • DoS attacks • Malicious insiders	• Network monitoring • Security inside design architecture • Policy enforcement • Secure key management • Data masking • Encryption • Obfuscation • Data classification	The user and the fog systems data may get compromised due to illegal file and database access if security parameters are not considered.

1.4.1 Current Observations and Future Security Prospects in Fog

Based on the study conducted here, this chapter presents the following observations and future recommendations in the area of fog security:

1. Data encryption is used to ensure data confidentiality, so its potential mechanisms can be used at different stages for securing data. For example, the Advanced Encryption Standard (AES) algorithm can be applied for data at rest, and the Secure Socket Layer (SSL) protocol can be used for data in transit. Integrity checks are to be mandated prior to and after the communications. Similarly, it is important to distinguish between the sensitive data and archival data such as public streaming videos. When encryption techniques are applied to such archived data, it may hamper the overall system performance due to constrained resources. So, strategies could be made in this direction in future work.
2. The frequently used data being stored in cache is prone to cache attacks, such as exposing cryptographic keys that might leak sensitive information. Therefore, strategies need to be devised involving hardware and software modifications that can prevent cache interference attacks.
3. A network is formed by connection of small devices. The data generated by one device may be small, but in a network with a number of communicating devices, the data becomes massive in quantity, creating difficulties in detecting anomalous activities as filtering each data packet can instigate more resource consumption. Anti-viruses, IDS, and firewalls can be used for efficacious network monitoring. In future research work, communications taking place at multiple levels can be monitored using rule matching patterns in artificial neural networks. Virtual private networks can also be established to isolate networks from external attacks.
4. The existing malware attacks, such as spyware, trojans, viruses, and worms may spread unwanted infections on sensitive data in the network. This highlights the need to implement an efficient cross-storage, light-weight detection service that can defend against these threats without compromising system performance.

TABLE 1.6
Abbreviations

Abbreviation	Description
AES	Advanced encryption standard
CPU	Central processing unit
DoS	Denial of service
HTTP	Hyper Text Transfer Protocol
IDS	Intrusion detection system
IoT	Internet of Things
P2P	Peer-to-peer
PKI	Public key infrastructure
SDN	Software defined network
SSL	Secure socket layer
TPA	Third party auditor

5. The fog paradigm has been widely implemented using wireless sensors and IoT devices. Techniques need to be developed that can provide solutions to packet sniffing and similar challenges. The mobility in these devices further complicates the implementation of security in communication as attackers get unprecedented freedom to intercept sensitive data. Therefore, more advanced Wi-Fi security algorithms, such as Wi-Fi protected access, wireless protocols such as 802.11a and 802.11g, and IDS for guarding in 5G heterogeneous mobile networks should be implemented for future correspondence.

Taking the threats posted in this chapter into consideration, along with directions for future contributions, a systematic system or model must be derived to provide secured storage, and computation and communication of data within constraints of resources. Such a model is envisaged to protect networks from potential damage and can avoid the occurrence of proactive vulnerabilities.

Table 1.6 presents the abbreviations used in this chapter.

REFERENCES

1. What is Fog Computing? - Definition from IoTAgenda. IoT Agenda. (2021). Retrieved 7 April 2021, from https://internetofthingsagenda.techtarget.com/definition/fog-computing-fogging.
2. Fog: A Powerful "Cloud Services" Gem. Rubyinside.com. (2021). Retrieved 7 April 2021, from http://www.rubyinside.com/fog-a-powerful-cloud-services-gem-3375.html.
3. Ni, L., Zhang, J., Jiang, C., Yan, C., & Yu, K. (2017). Resource Allocation Strategy in Fog Computing Based on Priced Timed Petri Nets. IEEE Internet of Things Journal, 4(5), 1216–1228. https://doi.org/10.1109/jiot.2017.2709814.
4. Salamone, S. (2021). Why Edge Computing Can Help IoT Reach Full Potential - RTInsights. RTInsights. Retrieved 3 February 2021, from https://www.rtinsights.com/why-edge-computing-can-help-iot-reach-full-potential/.

5. Bittencourt, L., Diaz-Montes, J., Buyya, R., Rana, O., & Parashar, M. (2017). Mobility-Aware Application Scheduling in Fog Computing. IEEE Cloud Computing, 4(2), 26–35. https://doi.org/10.1109/mcc.2017.27.
6. Bakker, R., Oppenheimer, P., Bakker, R., Story, B., Spade, J., & Bakker, R. et al. (2021). Perspectives - Page 27 of 47 - Cisco Blogs. Cisco Blogs. Retrieved 10 January 2021, from https://blogs.cisco.com/perspectives/page/27.
7. Wasim Akram, S., Rajesh, P., & Shama, S. (2018). A Review Report on Challenges and Opportunities of Edge, Fog and Cloud Computing by Employing IoT Technology. International Journal of Engineering & Technology, 7(3.29), 263. https://doi.org/10.14419/ijet.v7i3.29.18808.
8. Fog - The Ruby Cloud Services Library. Fog.io. (2020). Retrieved 10 September 2020, from http://fog.io/.
9. Mukherjee, M., Shu, L., & Wang, D. (2018). Survey of Fog Computing: Fundamental, Network Applications, and Research Challenges. IEEE Communications Surveys & Tutorials, 20(3), 1826–1857. https://doi.org/10.1109/comst.2018.2814571.
10. Mukherjee, M., Matam, R., Shu, L., Maglaras, L., Ferrag, M., Choudhury, N., & Kumar, V. (2017). Security and Privacy in Fog Computing: Challenges. IEEE Access, 5, 19293–19304. https://doi.org/10.1109/access.2017.2749422.
11. Ko, R., Jagadpramana, P., Mowbray, M., Pearson, S., Kirchberg, M., Liang, Q., & Lee, B. (2011). TrustCloud: A Framework for Accountability and Trust in Cloud Computing. 2011 IEEE World Congress on Services. https://doi.org/10.1109/services.2011.91.
12. Khan, K., & Malluhi, Q. (2010). Establishing Trust in Cloud Computing. IT Professional, 12(5), 20–27. https://doi.org/10.1109/mitp.2010.128.
13. Gupta, K. (2019). Shodhganga@INFLIBNET: Browsing Shodhganga. Shodhganga.inflibnet.ac.in. Retrieved 13 October 2020, from https://shodhganga.inflibnet.ac.in/browse.
14. Yi, S., Qin, Z., & Li, Q. (2015). Security and Privacy Issues of Fog Computing: A Survey. Wireless Algorithms, Systems, and Applications, 685–695. https://doi.org/10.1007/978-3-319-21837-3_67.
15. Damiani, E., di Vimercati, D., Paraboschi, S., Samarati, P., & Violante, F. (2002). A Reputation-Based Approach for Choosing Reliable Resources in Peer-to-Peer Networks. Proceedings of the 9th ACM Conference on Computer and Communications Security - CCS '02. https://doi.org/10.1145/586110.586138.
16. Law, Y. W., Palaniswami, M., Kounga, G., & Lo, A. (2013). WAKE: Key Management Scheme for Wide-Area Measurement Systems in Smart Grid. IEEE Communications Magazine, 51(1), 34–41. https://doi.org/10.1109/mcom.2013.6400436.
17. Abouhogail, R. (2011). New Multicast Authentication Protocol for Entrusted Members Using Advanced Encryption Standard. The Egyptian Journal of Remote Sensing and Space Science, 14(2), 121–128. https://doi.org/10.1016/j.ejrs.2011.11.003.
18. Ye, X., & Khoussainov, B. (2013). Fine-Grained Access Control for Cloud Computing. International Journal of Grid and Utility Computing, 4(2/3), 160. https://doi.org/10.1504/ijguc.2013.056252.
19. Homomorphic Encryption - Wikipedia. En.wikipedia.org. (2020). Retrieved 10 April 2020, from https://en.wikipedia.org/wiki/Homomorphic_encryption. Last edited on 23 May 2022.
20. What Is Homomorphic Encryption? - Definition from WhatIs.com. SearchSecurity. (2020). Retrieved 10 April 2020, from https://www.techtarget.com/searchsecurity/definition/homomorphic-encryption.
21. Matthew Green. (2020). A Few Thoughts on Cryptographic Engineering. Retrieved 10 April 2020, from https://blog.cryptographyengineering.com/author/matthewdgreen/page/12/.
22. Perera, C., Zaslavsky, A., Christen, P., & Georgakopoulos, D. (2014). Context Aware Computing for the Internet of Things: A Survey. IEEE Communications Surveys & Tutorials, 16(1), 414–454. https://doi.org/10.1109/surv.2013.042313.00197.

23. Saeed, W., & Hussain, R. (2014). Service Based Model Using Context Awareness for Ubiquitous Computing. International Journal of Computer Applications, 97(6), 21–22. https://doi.org/10.5120/17011-7286.
24. Tsugawa, M., Matsunaga, A., & Fortes, J. (2013). Cloud Computing Security: What Changes with Software-Defined Networking? Secure Cloud Computing, 77–93. https://doi.org/10.1007/978-1-4614-9278-8_4.
25. Stojmenovic, I., & Wen, S. (2014). The Fog Computing Paradigm: Scenarios and Security Issues. Annals of Computer Science and Information Systems. https://doi.org/10.15439/2014f503.
26. Liu, K., Ng, J., Lee, V., Son, S., & Stojmenovic, I. (2016). Cooperative Data Scheduling in Hybrid Vehicular Ad Hoc Networks: VANET as a Software Defined Network. IEEE/ACM Transactions on Networking, 24(3), 1759–1773. https://doi.org/10.1109/tnet.2015.2432804.
27. Fakeeh, K.A. (2016). Privacy and Security Problems in Fog Computing. Communications on Applied Electronics, 4(6), 1–7. https://doi.org/10.5120/cae2016652088.
28. Gennaro, R., Gentry, C., & Parno, B. (2010). Non-interactive Verifiable Computing: Outsourcing Computation to Untrusted Workers. Advances in Cryptology – CRYPTO 2010, 465–482. https://doi.org/10.1007/978-3-642-14623-7_25.
29. Parno, B., Howell, J., Gentry, C., & Raykova, M. (2013). Pinocchio: Nearly Practical Verifiable Computation. 2013 IEEE Symposium on Security and Privacy. https://doi.org/10.1109/sp.2013.47.
30. Song, D. X., Wagner, D., & Perrig, A. (2000). Practical Techniques for Searches on Encrypted Data. Proceeding 2000 IEEE Symposium on Security and Privacy. S&P 2000. https://doi.org/10.1109/secpri.2000.848445.
31. Mukherjee, B., Neupane, R., & Calyam, P. (2017). End-to-End IoT Security Middleware for Cloud-Fog Communication. 2017 IEEE 4th International Conference on Cyber Security and Cloud Computing (Cscloud). https://doi.org/10.1109/cscloud.2017.62.
32. Wang, H., Wang, Z., & Domingo-Ferrer, J. (2018). Anonymous and Secure Aggregation Scheme in Fog-based Public Cloud Computing. Future Generation Computer Systems, 78, 712–719. https://doi.org/10.1016/j.future.2017.02.032.
33. Zhang, P., Zhou, M., & Fortino, G. (2018). Security and Trust Issues in Fog Computing: A Survey. Future Generation Computer Systems, 88, 16–27. https://doi.org/10.1016/j.future.2018.05.008.
34. Khan, S., Parkinson, S., & Qin, Y. (2017). Fog Computing Security: A Review of Current Applications and Security Solutions. Journal of Cloud Computing, 6(1). https://doi.org/10.1186/s13677-017-0090-3.
35. Qiu, M., Kung, S., & Gai, K. (2020). Intelligent Security and Optimization in Edge/Fog Computing. Future Generation Computer Systems, 107, 1140–1142. https://doi.org/10.1016/j.future.2019.06.002.
36. Kukreja, V., Ahuja, S., & Singh, A. (2018). Measurement and Structural Model of Agile Software Development Critical Success Factors. International Journal of Engineering & Technology, 7(3), 1236. https://doi.org/10.14419/ijet.v7i3.12776.

2 An Elucidation for Machine Learning Algorithms Used in Healthcare

Veerpal Kaur
School of Computer Science and Engineering,
Lovely Professional University
Punjab, India

Rajpal Kaur
Department of Commerce and Education,
Maharaja Ganga Singh University
Rajasthan, India

CONTENTS

DOI: 10.1201/9781003143468-2

2.1 INTRODUCTION

Machine learning is the technology to make machines learn by training them with the available data, and once trained, the machines are tested for desirable results [1]. No doubt a machine can't cope with all the real-world data shortcomings, but it shows astonishing results when used in different application areas. The use of machine learning to train the systems has proven to be a successful technique that can be implemented by various other fields except computer science, such as earth science, applied sciences, healthcare, and any other field where there is need for data introspection. Machine learning is divided into four categories (cf Figure 2.1) [2], which are discussed in the subsequent sections.

2.1.1 Supervised Learning

The supervised learning algorithms are the ones that learn first and then they implement the results on the provided dataset. In a simple way, it can be said that the model is first learning itself and then on the basis of learnt skills, it processes the information provided. Examples of such algorithms are regression and classification algorithms.

2.1.2 Semi-supervised Learning

Semi-supervised learning, also known as pseudo learning, is a machine learning technique that uses the labeling method to train the model. It is a mixture of supervised and unsupervised machine learning techniques.

2.1.3 Unsupervised Learning

In this learning technique, the system does not have any labels to learn from, and hence, it is called unsupervised. It is used to detect hidden patterns available in the dataset and create data groups. Examples of such algorithms are clustering algorithms.

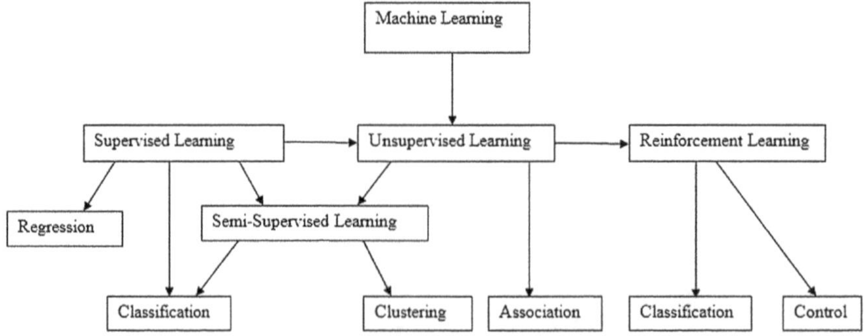

FIGURE 2.1 Listing of different machine learning algorithms.

2.1.4 Reinforcement Learning

This is the learning technique in which the system is allowed to use hit and trial and the system learns from its own experiences and errors. Examples of such learning techniques are Q-learning algorithms.

2.2 DECISION MAKING IN HEALTHCARE

Decision making is the most crucial part of healthcare system. If the decision-making system fails, fatal results may occur. Machine learning algorithms help medical staff to view these things far better. With prediction techniques, one can foresee the effect of disease; with classification, one can classify the different factors behind a disease occurrence; with clustering, one can group the people with same symptoms; and all these techniques help a research team to reach the ultimate stage of finding a cure to the disease.

2.2.1 Decision Making

Decision making is the procedure in which the doctor reaches a conclusion after going through the steps of meticulously analyzing data. Machines can greatly help doctors in this process by displaying visualization results of the huge data available.

2.2.2 Decision-making Architecture in Healthcare

Architecture should be followed to make effective use of the machines available by maintaining the records of the patients, accounts, medicines, and many other so as to support the whole architecture [7]. The data saved from previous patients will definitely aid doctors in making decisions to detect and diagnose a disease or other issues, and guide them on which medicines to prescribe to patients. So, the whole system will be influenced by the machines present in the diagnosis centers.

2.3 MACHINE LEARNING IN HEALTHCARE

The role of machine learning is much powerful than any other fields. Machine learning algorithms can be applied in studying and analyzing images, patient data retrieval, and predicting cases of a particular disease so that the support system will be ready for any kind of emergency. Without analyzing the forecasting image of the disease casualties, the whole healthcare system will be disrupted [8]. Machine learning helps foresee the impacts prior to their occurrence, and based on the estimates, the disease can be cured accordingly. The healthcare system is highly dependent on computer-aided machines that help diagnose and visualize the data interpretations to get a view of the data and its hidden patterns. Based on that, the patient will be prescribed specific medical procurements.

Currently, the healthcare system has collected immense information that has been compiled from hospitals directly and the databases of local pharmacies. From minor headache symptoms to major heart issues, all have been recorded and can be used to draw crucial and statistical analysis that can help to diagnose the disease and even generate instances of various other ailments to be cured.

2.4 MACHINE LEARNING SCOPE IN HEALTHCARE

The thorough study of hidden patterns, image diagnosis, and predictions from the data available makes using machine learning algorithms a must in the field of medical science. The treatment of many health issues can possibly be done using machine learning strategies. Given the algorithms used in machine learning are highly accurate and able to generate various visualizations for the given dataset, they can be put to use to get better results and a clear view of the current scenario for the disease diagnosis.

Patients looking for medical care have a trust level established by hospitals that have a proven track record of accurately diagnosing symptoms, and hence patients go for help to the same doctor again and again. So, in order to protect, and even cure, the patients, doctors have the responsibility to have a clear view of the patient's health. Using machine learning at different levels of the healthcare system makes the whole architecture relevant and reliable.

2.5 MACHINE LEARNING ALGORITHMS FOR HEALTHCARE

2.5.1 Support Vector Machine

The Support Vector Machine (SVM) was developed in the late 1990s. The SVM algorithm is used to perform sampling of bulky datasets and split the given dataset into variable groups. The SVM is mostly used to implement regression and classification problems [13].

2.5.2 Naive Bayes Classification

Based on the Bayesian theorem of probability, the Naive Bayes classification algorithm is used to characterize the data into different labels, and the given instances can be verified by classifying new data points into different groups. The Naive Bayes is used to classify a huge dataset [14].

2.5.3 Decision Tree

Decision tree is one of the finest and most used algorithms for the purpose of classification of data points. Labeling is used for the classification of data, and the root of the tree is the explicit condition that initializes the process and is further bifurcated into labeled classes that satisfy or dissatisfy the condition of the root node [15].

2.5.4 K-Nearest Neighbor

K-nearest neighbor (KNN) is a supervised classification algorithm that is quite easy to deploy and is simple to understand. In KNN, a centroid is taken and then the data points are calculated for the distance between the data point and centroid; the data point that is closest is grouped into the centroid set of points and the process is followed for all the data points. Hence, for a class labeled with a centroid, condition and distance is measured using Euclidean and Manhattan methods [16].

2.5.5 Fuzzy Logic

Fuzzy logic has been proved to be the best algorithm that uses fuzzy sets having values between 0 and 1. It is one of the most famous methods in artificial intelligence, and hence, has good accuracy [17].

2.5.6 CART

The Classification and Regression Tree is known as CART. In this strategy, the target is to get categorical and continuous values. These values are used to find data points in a tree, and the tree can be a regression tree or a classification tree [18].

2.6 PREDICTING SEVERAL DISEASES USING MACHINE LEARNING

2.6.1 Machine Learning in Heart Diagnosis

A huge intake of unhealthy food, like junk food, that has more cholesterol and other fats harms the heart to a great extent. The precision is of utmost importance in the process of analyzing heart-related issues. The data presented in the Table 2.1 has been taken from various sources and precisely reflects the machine learning usage and its impacts. The SVM algorithm has been compared with J48 and Naive Bayes, and the results reflected 94.60% precision with SVM and 74% precision with Naive Bayes [18].

2.6.2 Machine Learning in Diabetes Diagnosis

Diabetic issues have been profound among old age people nowadays; hence, viable machine learning approaches need to be among the methods used to diagnose diabetes among patients. The dataset implemented the Naive Bayes and Decision Tree algorithms get the precision percentage. A higher proportion of accuracy is reflected by Naive Bayes than decision tree [21].

2.6.3 Machine Learning in Cancer Diagnosis

The dataset used from University of California Irvine (UCI) repository identifies a better algorithm for the diagnosis of cancer. The J48 algorithm seems quite good at its job with a 94.2% precision rate in comparison to Naive Bayes with 83% [23]. Various other algorithms have also been proposed for breast cancer treatment, such as KNN, decision tree, and SVM. Among these, SVM has stolen the show by getting the highest accuracy rate of 96% [24].

TABLE 2.1
Use of Machine Learning in Healthcare

Machine Learning Model Used	For Disease Category	Dataset Taken from	Precision
SVM	Heart	UCI	85.03%
J48 [27]			84.35%
Naive Bayes [28]	Heart	Diabetic Research Institute, Chennai	86.41%
SVM [29]	Diabetes	UCI	78%
Naive Bayes [30]	Diabetes- type2	Different sectors of society in India	95%
J48 [31]	Breast cancer	UCI	98.14%
Decision tree [32]	Breast cancer	Swami Vivekananda Diagnostic Centre	97%
Decision tree (DT) + SVM [33]	Breast cancer	UCI	91%
CART	Breast cancer	University of Wisconsin Hospital	92.42%
Naive Bayes [34]			97.42%

2.6.4 Machine Learning in Thyroid Diagnosis

Fuzzy logic, SVM, and decision tree classification algorithms were deployed onto the dataset taken from UCI repository to get a clear view of the accuracy of the chosen algorithms. The implementation results reflect that fuzzy logic has been the best among the category [26].

Table 2.1 depicts the diagnosis approaches using machine learning techniques with their precision rate.

2.7 RELATED WORK

A rigorous literature review has been conducted to dive deep into the work already done, and a systematic tabular representation of it appears in Table 2.2. The table includes reference paper numbers with their titles and advantages, as well as their future scopes.

2.8 CONCLUSION

To conclude, this chapter reveals machine learning approaches that are being used in healthcare and that can be used in near future to facilitate the decision-making system in the medical science field. Machine learning approaches are prominently known for their accuracy; hence, they can be incorporated into healthcare to classify patient disease categories and can provide a better visualization of the methods that can be followed to cure diseases in a better way. Image recognition and their hidden patterns can be studied using machine learning algorithms, and an amalgamation of machine learning with healthcare technological advancements will lead to medical science that is par excellence. The future of healthcare is predicted to be a brighter one with the use of fifth-generation techniques, and hence will surely lead to a new world of healthy living.

TABLE 2.2
Literature Review of Machine Learning in Healthcare

Reference No.	Title of the Study	Benefits	Future Work
35	Applications of Machine Learning in the Field of Medical Care	Excelled in the previous system for proper medical distribution and resources. The paper presented the techniques of computer sciences such as artificial intelligence to explore the benefits of the current technology into health care. The paper also dived deep into the historical study of machine learning and its applications in medicinal field.	The study will focus on the human power and machine power to create a balance between both of them.
36	Breast Cancer Detection Using Machine Learning Algorithms	Prediction approaches using machine learning are presented in the paper for breast cancer. With the best accuracy, precision, and F1 score, KNN has been proven to be the best algorithm for the said purpose in comparison to other algorithms.	Other supervised machine learning algorithms are yet to be explored for breast cancer early diagnosis and prognosis.
37	A Literature Review on Machine Learning Based Medical Information Retrieval Systems	The paper presents a deep exploration into artificial intelligence used in medical field. The information retrieval (IR) algorithms for managing the big data related to medical field are also discussed.	The need for big data still exists to train the machine learning model used for medical information retrieval.
38	Medical Imaging Using Machine Learning and Deep Learning Algorithms: A Review	A standard dataset has been used to predict the required diseases using the images. Both supervised and unsupervised techniques have been used for the purpose. The search approach for the best algorithm in terms of accuracy is proposed in the paper.	Future work will look for such algorithms that may help the medical image inferences make crucial decisions.
39	Machine Learning for Improved Diagnosis and Prognosis in Healthcare	An inference system using the Bayesian approach has been proposed for diagnosing Alzheimer's disease,	The search continues for a better classification algorithm for classifying cell images to detect the breast cancer at an early stage. Significant work needs to be done to arrange huge datasets rich in figures for the optimum accuracy of diagnosis.

(Continued)

TABLE 2.2 (*Continued*)
Literature Review of Machine Learning in Healthcare

Reference No.	Title of the Study	Benefits	Future Work
40	Heart Disease Identification Method Using Machine Learning Classification in E-Healthcare	To solve the feature selection issues, a fast, conditional, novel feature selection algorithm has been proposed. The proposed system ensures the accuracy and reduction in execution time for the given classification algorithm. The proposed approach is working optimally with the Support Vector Machine (SVM) classifier to detect heart diseases.	Future work of the study will focus on the optimization of the proposed approach.
41	Prediction of Diabetes Using Machine Learning Algorithms in Healthcare	The presented work focuses on the prediction algorithms in machine learning that best suit the prediction of diabetes.	To get better accuracy, the future work will emphasize the integration of other machine learning algorithms with the proposed approach. Testing the proposed approach using big datasets that are pre-processed will help geta clearer view of the accuracy of the proposed approach.
42	Deep Learning for Health Informatics: Recent Trends and Future Directions	The paper studies the deep learning domain of artificial intelligence and its application in helping the healthcare sector with decision making.	
43	Comparative Study of Machine Learning Algorithms for Breast Cancer Detection and Diagnosis	The study presents the comparison between three prominent machine learning algorithms for detecting and diagnosing breast cancer in patients. The algorithms used in the study are SVM, Random Forest (RF), and Bayesian Network (BN). The study shows that SVM has the highest prediction of accuracy as well as precision.	
44	Architecture of Smart Health Care System Using Artificial Intelligence	The proposed system has the capability to find hidden patterns in the data given to the model, and hence helps medical officials get better insight into the data and help them make crucial decisions, saving time and effort.	

(*Continued*)

TABLE 2.2 (*Continued*)
Literature Review of Machine Learning in Healthcare

Reference No.	Title of the Study	Benefits	Future Work
45	Prediction of Cardiovascular Disease Using Machine Learning Algorithms	The paper presents the prediction of heart-related issues in the given patient's data. It helps in the diagnosis steps of the process.	The future study will focus on the hybrid of the algorithms discussed to get better performance in most areas of evaluation.
46	Blockchain and Machine Learning in Health Care and Management	The study focuses on the blockchain technology along with machine learning techniques that can be used in the healthcare sector to help manage data more securely and effectively.	The future work of the study will discuss implementing Internet of Things (IoT), along with artificial intelligence, in a healthcare system.
47	Predictive Analytics in Health Care Using Machine Learning Tools and Techniques	The paper focuses on the electronic maintenance of patient-related data, the prediction and diagnosis of various diseases, and the need of a suitable machine learning model.	Highly unstructured distributed and constantly changing datasets are the key challenges yet to be focused.
48	Applying Best Machine Learning Algorithms for Breast Cancer Prediction and Classification	The proposed study has implemented four different machine learning algorithms to predict breast cancer in patients.	More rigorous studies on machine learning techniques to get a hybrid approach with deep learning algorithms will be interesting to see.
49	Machine Learning Model for Breast Cancer Prediction	The convolution neural network used to detect breast cancer has been presented in the study. The results show better accuracy rates of the proposed approach. A systematic calculation has also been implemented to draw the results from the proposed technique.	
50	Machine Learning Based System for Prediction of Breast Cancer Severity	Artificial Neural Network (ANN), KNN, Binary Support Vector Machine (Binary SVM), and Decision Tree (DT) are the analyzed algorithms for getting a better computer-aided diagnosis for predicting cancer in patients and helping reduce death tolls.	The optimization of the proposed work is yet to be addressed.
51	Regression Analysis of COVID-19 Using Machine Learning Algorithms	The study presents a deep analysis of COVID-19 among various states using the regression approach of machine learning.	Predicting the disease using a more accurate and precision-efficient algorithm from machine learning domain is to be addressed.

REFERENCES

1. T. Mitchell, "Machine Learning," McGraw Hill. p. 2, 1997.
2. A. Mishra, A. Shukla, "From Machine Learning to Deep Learning Trends and Challenges," CSI Communications, December 2018.
3. N. M. Allix, "Epistemology and Knowledge Management Concepts and Practices." Journal of Knowledge Management Practice, vol. 4(1), pp. 1–24, 2003.
4. A. Mathur, G. P. Moschis, "Socialization Influences on Preparation for Later Life." Journal of Marketing Practice: Applied Marketing Science, vol. 5, pp. 163–176, 2007.
5. L. P. Kaelbling, M. L. Littman, A. W. Moore, "Reinforcement Learning: A Survey," Journal of Artificial Intelligence Research, vol. 4, pp. 237–285, 1996.
6. T. G. Thompson, D. J. Brailer, "The Decade of Health Information Technology: Delivering Consumer-Centric and Information-Rich Health Care," US Department of Health and Human Services, 2004.
7. K. Rajalakshmi, S. C. Mohan, S. D. Babu, "Decision Support System in Healthcare Industry," International Journal of Computer Applications, vol. 26(9), pp. 42–44, 2013.
8. R. Bhardwaj, A. R. Nambiar, "A Study of Machine Learning in Healthcare," IEEE 41st Annual Computer Software and Applications Conference, 2017.
9. M. Hauskrecht, S. Visweswaran, G. Cooper, G. Clermont, "Clinical Alerting of Unusual Care That Is Based on Machine Learning from Past EMR Data," 2015.
10. M. Kohn, "Real World Data and Clinical Decision Support," Elsevier, 2009.
11. J. Sukanya, "Applications of Big Data Analytics and Machine Learning Techniques in Health Care Sectors," International Journal of Engineering and Computer Science, vol. 6, pp. 21963–21967, 2017.
12. K. P. Murphy, "Machine Learning: A Probabilistic Perspective," The MIT Press, 2012.
13. A. Hazra, S. K. Mandal, A. Gupta, "Study and Analysis of Breast Cancer Cell Detection Using Naïve Bayes, SVM and Ensemble Algorithms," International Journal of Computer Applications, vol. 145(2), pp. 39–45, 2016.
14. P. Sharma, A. P. R. Bhartiya, "Implementation of Decision Tree Algorithm to Analysis the Performance," vol. 10, pp. 861–864, 2012.
15. C. M. Bishop, "Neural Networks for Pattern Recognition," Oxford University, 1995.
16. H. Zimmermann, "Fuzzy Set Theory and Its Applications," Kluwer Academic Publishers, 2001.
17. A. Hazra, S. K. Mandal, A. Gupta, A. Mukherjee, A. Mukherjee, "Heart Disease Diagnosis and Prediction Using Machine Learning and Data Mining Techniques: A Review," Advances in Computational Sciences and Technology, vol. 10, pp. 2137–2159, 2017.
18. G. Parthiban, S. K. Srivatsa, "Applying Machine Learning Methods in Diagnosing Heart Disease for Diabetic Patients," International Journal of Applied Information Systems, vol. 3, pp. 25–30, 2012.
19. A. F. Otoom, E. E. Abdallah, Y. Kilani, A. Kefaye, "Effective Diagnosis and Monitoring of Heart Disease," International Journal of Software Engineering and Its Applications, vol. 9, pp. 143–156, 2015.
20. A. Iyer, S. Jeyalatha, R. Sumbaly, "Diagnosis of Diabetes Using Classification Mining Techniques," International Journal of Data Mining & Knowledge Management Process (IJDKP), vol. 5, pp. 1–14, 2015.
21. S. K. Sen, S. Dash, "Application of Meta Learning Algorithms for the Prediction of Diabetes Disease," International Journal of Advance Research in Computer Science and Management Studies, vol. 2, pp. 396–401, 2014.
22. K. Williams, P. A. Idowu, J. A. Balogun, A. I. Oluwaranti, "Breast Cancer Risk Prediction Using Data Mining Classification Techniques," Transactions on Networks and Communications, vol. 2, pp. 1–11, 2015.

23. Z. K. Senturk, R. Kara, "Breast Cancer Diagnosis via Data Mining: Performance Analysis of Seven Different Algorithms," Computer Science & Engineering, vol. 1, pp. 1–10, 2014.
24. J. Majali, R. Niranjan, V. Phatak, O. Tadakhe, "Data Mining Techniques for Diagnosis and Prognosis of Cancer," International Journal of Advanced Research in Computer and Communication Engineering, vol. 3, pp. 613–616, 2015.
25. E. I. Papageorgiou, N. I. Papandrianos, D. J. Apostolopoulos, P. J. Vassilakos, "Fuzzy Cognitive Map Based Decision Support System for Thyroid Diagnosis Management," International Conference on Fuzzy Systems, pp. 1204–1211, 2008.
26. V. Chaurasia, S. Pal, "Data Mining Approach to Detect Heart Disease," International Journal of Advanced Computer Science and Information Technology, vol. 2, pp. 56–66, 2018.
27. K. Vembandasamy, R. Sasipriya, E. Deepa, "Heart Diseases Detection Using Naive Bayes Algorithm," vol. 2, pp. 441–444, 2015.
28. V. A. Kumari, R. Chitra, "Classification of Diabetes Disease Using Support Vector Machine," International Journal of Engineering Research and Applications, vol. 3, pp. 1797–1801, 2013.
29. A. Sarwar, V. Sharma, "Intelligent Naïve Bayes Approach to Diagnose Diabetes Type-2. Special Issue," International Journal of Computer Applications and Challenges in Networking, Intelligence and Computing Technologies, vol. 3, pp. 14–16, 2012.
30. S. S. Shrivastavat, A. Sant, "An Overview on Data Mining Approach on Breast Cancer Data," International Journal of Advanced Computer Research, vol. 3, pp. 256–262, 2013.
31. E. Venkatesan, T. Velmurugan, "Performance Analysis of Decision Tree Algorithms for Breast Cancer Classification," Indian Journal of Science and Technology, vol. 8, pp. 1–8, 2015.
32. K. Sivakami, N. Saraswathi, "Mining Big Data: Breast Cancer Prediction Using DT-SVM Hybrid Model," International Journal of Scientific Engineering and Applied Science, vol. 1, pp. 418–429, 2015.
33. S. S. Shajahaan, S. Shanthi, V. ManoChitra, "Application of Data Mining Techniques to Model Breast Cancer Data," International Journal of Emerging Technology and Advanced Engineering, vol. 3, pp. 362–369, 2013.
34. M. R. N. Kousarrizi, F. Seiti, M. Teshnehlab, "An Experimental Comparative Study on Thyroid Disease Diagnosis Based on Feature Subset Selection and Classification," International Journal of Electrical & Computer Sciences, vol. 1, pp. 13–19, 2012.
35. H. Dou, "Applications of Machine Learning in the Field of Medical Care," 34th Youth Academic Annual Conference of Chinese Association of Automation (YAC), 2019.
36. S. Sharma, A. Aggarwal, T. Choudhury, "Breast Cancer Detection Using Machine Learning Algorithms," International Conference on Computational Techniques, Electronics and Mechanical Systems (CTEMS), 2018.
37. A. Gudivada, N. Tabrizi, "A Literature Review on Machine Learning Based Medical Information Retrieval Systems," IEEE Symposium Series on Computational Intelligence (SSCI), 2018.
38. J. Latif, C. Xiao, A. Imran, S. Tu, "Medical Imaging Using Machine Learning and Deep Learning Algorithms: A Review," International Conference on Computing, Mathematics and Engineering Technologies, IEEE, 2019.
39. N. G. Maity, Dr. S. Das, "Machine Learning for Improved Diagnosis and Prognosis in Healthcare," IEEE Aerospace Conference, 2017.
40. J. P. Li, A. UlHaq, S. UdDin, J. Khan, A. Khan, A. Saboor, "Heart Disease Identification Method Using Machine Learning Classification in E-Healthcare," IEEE Access, vol. 8, pp. 107562–107582, 2020.

41. M. A. Sarwar, N. Kamal, W. Hamid, M. A. Shah, "Prediction of Diabetes Using Machine Learning Algorithms in Healthcare," 24th International Conference on Automation and Computing (ICAC), IEEE, 2018.
42. S. Srivastava, S. Soman, A. Rai, P. K. Srivastava, "Deep Learning for Health Informatics: Recent Trends and Future Directions," International Conference on Advances in Computing, Communications and Informatics, IEEE, 2017.
43. D. Bazazeh, R. Shubair, "Comparative Study of Machine Learning Algorithms for Breast Cancer Detection and Diagnosis," 5th International Conference on Electronic Devices, Systems and Applications, IEEE, 2016.
44. M. M. Kamruzzaman, "Architecture of Smart Health Care System Using Artificial Intelligence," IEEE International Conference on Multimedia & Expo Workshops, 2020.
45. D. G. Kumar, K. Arumugaraj, V. Mareeswari, "Prediction of Cardiovascular Disease Using Machine Learning Algorithms," International Conference on Current Trends towards Converging Technologies, IEEE, 2018.
46. S. Jain, A. Anand, A. Gupta, K. Awasthi, S. Gujrati, J. Channegowda, "Blockchain and Machine Learning in Health Care and Management," International Conference on Mainstreaming Block Chain Implementation, IEEE, 2020.
47. B. Nithya, Dr. V. Ilango, "Predictive Analytics in Health Care Using Machine Learning Tools and Techniques," International Conference on Intelligent Computing and Control Systems, IEEE, 2017.
48. Y. Khourdifi, M. Bahaj, "Applying Best Machine Learning Algorithms for Breast Cancer Prediction and Classification," International Conference on Electronics, Control, Optimization and Computer Science, IEEE, 2018.
49. A. Gupta, D. Kaushik, M. Garg, A. Verma, "Machine Learning model for Breast Cancer Prediction," Fourth International Conference on I-SMAC (IoT in Social, Mobile, Analytics and Cloud), IEEE, 2020.
50. S. Laghmati, A. Tmiri, B. Cherradi, "Machine Learning Based System for Prediction of Breast Cancer Severity," International Conference on Wireless Networks and Mobile Communications, IEEE, 2019.
51. E. Gambhir, R. Jain, A. Gupta, "Regression Analysis of COVID-19 Using Machine Learning Algorithms," International Conference on Smart Electronics and Communication, 2020.

3 Tea Vending Machine from Extracts of Natural Tea Leaves and Other Ingredients

IoT and Artificial Intelligence Enabled

Neha Sharma, Ram Kumar Ketti Ramachandran, Huma Naz, and Rishabh Sharma
Chitkara University Institute of Engineering & Technology, Chitkara University
Punjab, India

CONTENTS

DOI: 10.1201/9781003143468-3

3.1 INTRODUCTION

As technology advances every day, vending machines provide simple as well as complex services to users according to their needs. The design, automation, and technology of vending machines are changed as per the users' needs. Various products and services are offered by these automated machines, and they all fall under vending machines [1]. The very first vending machine was developed only for individual product delivery, such as for newspapers, cigarettes, beverages, and so on. Other types of services, such as transportation tickets and vehicle washing [2], are categorized as a payment for one service. These machines are commonly installed in crowded or busy places, such as bus/train stations, schools, airports, companies, shopping malls, and universities [3]. Users can purchase products without wasting time and can quickly make cashless payments [4, 5]. Some places offer 24 × 7 service, which explains why vending machines have gained so much of popularity and have led to large profits for vending machine owners [6]. All around the world, beverages like tea and coffee have become a part of life as well as an everyday routine. Hence, it is a very big market on a daily basis. The global intelligent vending machine market already reached a market value of more than $5,000 million in 2017 and is expected to reach more than $15 billion by 2025. It is estimated to grow at a compound annual growth rate (CAGR) of 15.3%.

In general, pure tea is made up of tea leaves (*Camellia sinensis*) that do not contain any additives. However, flavored/packed teas have many preservatives and artificial elements, and tea powders added to them contain possible pesticides, artificial sweeteners, coloring agents, and other harmful chemicals to keep them fresh and increase their longevity. Colored tea can legally contain, but in practice rarely contains, artificial colors that are openly advertised as such. However, there has been a persistent problem with people adding unreported colorings to tea, often as a way to "stretch" low-quality tea further. These colorings can include unsafe ingredients. Some undisclosed ingredients in teas include coal tar dye, indigo, soapstone, Prussian blue, and gypsum. Even renowned tea brands are suspected of overusing additives. Concentrated supplements often use lower-quality ingredients, and have a higher risk of contamination with heavy metals or other toxins, whereas high-quality loose-leaf teas are less likely to contain contaminants. Most of the common additives to tea are safe, but there are occasional concerns with contaminants or other unlabeled additives [7]. Some additives, such as citric acid, unwanted sweeteners, or natural and artificial flavors, can produce undesirable flavor characteristics and/or be used to mask low-quality ingredients. Products marketed primarily for "health" reasons, especially those containing concentrated extracts or supplements, pose the highest risk of causing harm.

Nowadays, vending machines are connected to the Internet that can send and receive signals remotely [8]. Over the last decade machine-to machine (M2M) communication has been happening [9]. Vending machines are refilled using telemetry for the daily operations, whereas cashless payment is used just to boost sales and increase consumer convenience. Due to high competition in the vending market in every country, small and big markets are fragmented everywhere [10, 11].

3.2 METHODOLOGY

Methodology includes planning, selection, design, and the complete setup of the final product (Figure 3.1).

In this methodology, we first plan the project; for this, a proper survey is conducted, and the problem is identified. To solve the problem, we then research a few existing systems. For better results with the tea, ingredients are added so that a pure and fresh composition is available for the grinding and mixing unit when added to the existing machine. The unit will crush all the ingredients and produce fresh tea. After that, the modification system is connected to the cloud, where it will

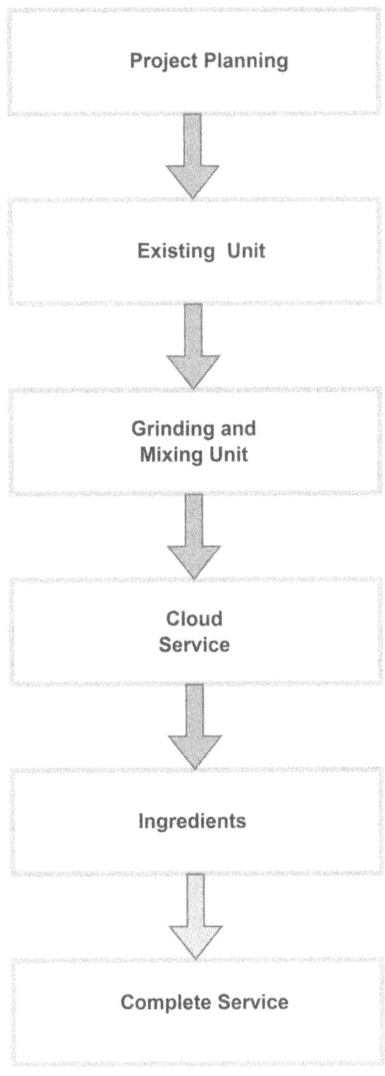

FIGURE 3.1 Stages of tea vending unit.

calculate the data added by users and give more intelligent solutions to users. At the same time, it will collect all the user data and create a database. Ingredients will help to reduce quantity of the tea; it will be natural as well as will cure many diseases. Finally, a complete solution in the form of a complete unit is proposed. It will include all the ingredients and provide the desired herbal tea according to the consumer's wishes.

3.3 EXISTING SYSTEM AND MECHANISM

The vending machine is made up of brushes, metals and motors [12]. The goal is to deliver beverages and snacks to users by operating each block on the motor-based mechanism. To place an order, a person has to make payment first after which they can collect a printable token and also avail other options; for example, an espresso. Nowadays, a variety of drinks, including latte, hot water, milk, and green coffee, are also available in the smart vending machine with the bolted position. To serve things to the client, a person can open the container and take out papers. The same mechanism can also facilitate other vending machines used for food delivery, snacks, paper vending as well as beverages vending, and a continuous health monitoring status and timely dosage of patients. The proposed system helps healthcare officials to view the effects of medication and remotely monitor their patients.

3.4 BIOGRAPHY OF THE VENDING MACHINE

The first vending machine was invented by Hero of Alexandria, one of the famous designers and mathematicians in Roman Egypt. The machine worked using coins, also known as metal currency. When a coin was inserted via a slot at the top of the machine, it fell upon a container attached to a lever and released holy water. The container continued to tilt with the weight of the coin until it fell off, and the valve would close. In 1822, Richard Carlile manufactured a paper-administering machine for a book shop. Simeon Denham, an Englishman, used this working principle for the collection of warm water. It was the first vending machine tested in the hospital in 1867 [13, 14].

3.5 EXISTING SYSTEM

With the advancements in technology, vending machines can now sell simple to complex products to the users opting different services. A large amount of data is collected every day that connects such vending machines to the internet. And now tea and coffee vending machines are also using the Internet of Things (IoT) devices. As per the inputs received from the users, the desired beverage is made. All around the world, tea and coffee play a very important role in everyone's life. Working people, such as multinational corporation (MNC) employees and manufacturers, want to have their favorite beverages at their workplace. Although an MNC can afford beverages from the cafeteria, small organizations cannot. In the case of small organizations, beverage requirements are fulfilled through the street vendors every day, but the quality of their products can be questionable, as they use tap water to prepare such beverages, which can drastically affect their quality. Another issue is cleanliness because the utensils used to prepare these beverages cannot always be trusted. Moreover, different people have different preferences for the beverages; for example, some people take strong sugar, some take less sugar, and others consume beverages without even using sugar. Meeting these requirements can be a bit difficult when one relies on street vendors for beverages [15]. The major issue is the time taken to dispense them in cups because a single minute is very precious for the working people and it takes more than 15 minutes just to reach consumers. Here, the use of tea and coffee vending machines is the complete solution to the problem [16]. Tea and coffee vending machines are already available in the market, but their price and size make them unaffordable to small organizations. The advancements in vending machines are being made continuously due to which these have been improving logically as well as in their facility as compared to earlier vending machines. These advancements have added many important features to the vending machines such as sensor availability, web and camera availability, lower power consumption rate, lower costs, more computerized contacts, and radio frequency identification (RFID). The advanced vending machines provide the advantage to the consumers with [17] different experiences and decreases working expense through wireless communication. Some latest technologies such as artificial intelligence (AI), sensors, and camera features are also incorporated in the vending machines to enhance the web-based social networking user experience.

3.6 LITERATURE SURVEY

Nourishment items provide nourishment and refreshment to people. But when they get involved with the contribution element, it results in performance and technology-based improvement in the existing system; the same is the case with vending machines. An outline methodical audit assessment is carried out on the basis of smart evaluation technology and present procedure, and survey results are collected [19]. An evaluation of the implementation factors includes the type of apparatus used for the experiment, the items accessed, the openness of machine, the area of the study, the size of partitioning, advancements in the items, and the logic used to train [20]. The result of the audit is that logical practices are used only 22%, smart vending machine accessibility is 39%, the assessed cost is 48%, advancement is 52%, 70% is the segment assessed, accessibility of item assessed is 91%, and built-up invigorating effect criterion is 96% assessed from the investigated articles. It is found that from the numerous articles, only 87% gave resolutions of the vending machine condition [21]. In an analysis between these existing automatic systems, it is difficult to achieve results; however, it is easy to agree with the dependable vending machine apparatus [22, 23]. Coffee beverage is offered by the IoT-based vending machine to the clients.

Further, it also includes a few items, such as hot coffee/tea, espresso, latte, titbits, and cool drink beverages, as a refreshment to the users [24]. Earlier, all the payments are currency-based, but the existing systems use RFID-based technology and smart cards. This framework identifies the user access by reading the card using radio frequency. After identifying the product [25] and cost via RFID, database is updated and the desired output (product) is dispensed. If any problem or issue arises, the RFID number needs to be fixed; the presence of any magnetize material gives an error to the RFID card and it stops working. A small chip is placed in the smart system reader of an online payment-based coffee vending system. According to the necessity of the users per day, the quantity of beverages in the cup is customized [26]. As desired by the user, each cup is dispensed by the representative, and if a user desires to have more quantity of beverages than the usual one, then the sum can easily be cut from the compensation account. While using vending machines, human assistance is not required to serve customers. Simple steps are performed by the consumers just to get the diamond and platinum jewelry from the vending machine after inserting the card [27]. A controlled system is used just to perform payment, communication, and product extraction. Food and nonfood items and other convenience products are sold by the different markets using these vending machines. In the past years, selling hot and cold drinks was the main business, but nowadays, due to the variety of food and other services, the market for the vending machine and its products is increasing constantly. The vending machine is recognized [29] differently by different countries and their vending associations. Vending is defined for the small article or product, operated by the coin, and the small article or product is sold by using a vending machine (Oxford University Press 1994, p. 568) [30]. In Japan, large quantities of products are also sold through vending machines; for example, a customer can buy ten kilogram rice bags from the vending machines. Credit card is used in the slot of card in the vending machine to pay for goods and services.

According to the National Automatic Merchandising Association (NAMA) in the United States, vend is the delivery of a single unit of merchandise (NAMA, www.vending.org, 2005). For vending, there is a slogan in the United States, "Coffee, Candy, Cola" (Cola" o. V. 1999 Coffee, Candy, Cola, p. 17ff) [31]. The term "Coffee" indicates hot drinks like hot chocolate, soup, and coffee. "Candy" illustrates the sweets, and "Cola" is used for different types of carbonated soft drinks. The early stages of the vending industry started with the concept of the 4Cs: cigarettes, coffee, candy, and cola. Once the consumer rate increased, this industry grew to 7Cs from the 4Cs concept. These 7Cs are candy, cold drinks, coffee, cigarettes, canned drinks, cold cups, chips (NAMA, www.vending.org/nama_vision/index.php?page=definitions, 2005) [32]. In Europe, extensive products are involved in the vending machine (EVA, European Vending Association) (EVA, www.eva.be, 2005) [33]. In Germany, daily use products, including food items, drinks, and and nonfood products, are sold using vending machines.

According to the Australian Association, vending is used to sell all products, including drinks, food, parking tickets, as well as photos. Apart from these products, pin balls, slot machines, lockers, telephones, and copying machines are also included (ÖVV, Österreichische Verkaufsautomaten Vereinigung, 2005) [34]. The vending machine is treated as a store in which all the products are stored for the retail trade industry. It follows an automatic procedure of selling the product. A customer just needs to select the item, pay for it, and collect the item. In the United States, for the selection of the product, the machine products appear behind the glass so that customers can select the product easily. A spiral mechanism facilitates easy delivery of the product to the customer. And the product is dispensed at the bottom of the machine for delivery (NAMA, www.vending.org, 2005) [35].

Therefore, vending is the term that is used to provide all the services and ease to the customers by collecting cash in different formats, such as cash, credit, and by text messaging as well as electronically. There are a few tea and coffee vending machines with simple working, such as the Morphy Richard Tea Maker and Philips HD 7450 [36]. These vending machines produce a very limited amount of beverages. The working principle of these vending machines is based on the requirement of users because they use plastic containers in which the tea and coffee powder is already mixed. The water reservoir is connected to the heater just to mix the tea and coffee powder. According to the user's desired quantity, the beverage is served to the user once the water gets heated and mixed with the powder. Generally, these vending machines are used in hotels, offices, and cafeterias, and all the operations performed in the vending machine are further controlled by the electronic and mechanical system [37]. The interface is already decided in the specific part of the machine where the user has to give the desired input just to get the beverage. After complete registration of the user's input, the timing circuit provides all the control while mixing the powder and adding the water, and the beverage is delivered to the main cup with the flow.

Xie et al., 2017 [38] have proposed a routing implementation of efficient energy that introduces the mobility of the objects with Wireless Sensor Networks (WSNs) for collection of information of mobile data. The term used for sensing moveable sensors is known as Mobile Data Collectors (MDCs). MDC starts the movement of data from its initial place-based station, transfers the data periodically to all the stations, and finally returns to the based station again. This whole movement of data from a station to another is decided using Floyd Warshall's complete graph algorithm.

Chandra et al., 2017 [39] discuss techniques based on data-driven results that help improve increasing productivity yields in the agriculture field. Moreover, this technology has a high rate of manual information gathering in bare implementation and gives a limited connectivity result. The authors present an uninterrupted IoT platform known as FarmBeats for agricultural solutions to collect data from the various drones, cameras, and sensors. FarmBeats is a method that helps monitor crop through the internet and has been implemented in 2017 in the two U.S. farms.

Miqdad et al., 2017 [40] developed a real-time low-cost WSNs monitoring system for sensing building space. The monitoring system collects data such as temperature, illumination, quality of air, and humidity. This data is taken from all corners of the building to balance the amount of energy and client comfort. In the experiment, the authors used the NodeMCU module with a DHT11 sensor for measuring humidity and temperature. The authors used a 6-volt power supply source to provide supply to the module NodeMCU.

AlSkaif et al., 2017 [41] compared environment monitoring of smart cities in four specific areas, that is smart building/house, smart animal, agriculture farming, and municipal resilience. The main purpose of the research is to make a comparative analysis of Multiple Access Control (MAC) protocol and energy utilization in low-rate data Wireless Multimedia Sensor Networks (WMSNs). Comparatively, the analysis includes the multiclass traffic model to analyze the utilization of energy of MAC protocol in low data rates. Modieginyane et al., 2018 [42] reviewed the challenges in the application faced with environment monitoring by WSNs.

Sethi and Sahoo [43] have proposed an application based on Health Monitoring System (HMS). The application is purely based on WSNs, which help monitor medical information or records in real time. Apart from industrial uses, the application is used to collect data from adult patients and intensive care units. In the experiment, the authors used ZigBee, Wi-Fi to transfer the data to the application.

Kaur et al., 2020 [44] discuss the effective role of IoT in the healthcare devices sector. IoT-based devices monitor the patient's health status and track their activities according to the current time and situation. Further, a viewpoint is given in the field of electronics and technology [45] which is related to the Internet of Healthcare Things (IoHTs). Finally, the authors discuss the benefits of IoHTs in the medical sector.

3.7 OBJECTIVES

- Proposed a new tea vending machine that will prepare tea from the fresh tea leaves on the spot without adding any artificial preservatives to give tea lovers a healthy tea.
- The cleanly washed fresh leaves are kept in green/dry format in the transparent jar from which tea powder is processed and directly sent to the existing tea vending machine.
- This type of machine does not exist in the market, and it will give a new experience to the end users.

3.8 ARCHITECTURE

Figure 3.2 gives the overview of the proposed machine; the functional units are described here.

1. **Transparent Funnel:** This can store 250 grams of fresh/dry tea leaves; this container will have an automatic open and close window to release a specific amount of the tea leaves. The operator can set the amount of leaves that are released at a time. It is fast since the quantity of the leaves taken is very limited and we have fast grinders and driers to improve the overall speed of the process.

2. **Grinding Unit:** This unit will crush and grind the leaves to the standard size of tea particles; it is a four-layer unit that will grind, crush, and filter the tea leaves. A soundproof system is optional to operate this grinder.
3. **Dryer:** It will dry all the leaves and remove moisture at the required temperature. There will be a cooling unit to cool down the dried particles. This entire unit is optional since there will be two modes of operations: fresh leaves or dry leaves. This unit is only functional when fresh tea leaves are fed.
4. **Tea Particles:** After grinding and drying, we will get tea. The granules of 0.33 mm size were found to show 98% dissolution rate as compared to 68% in case of 1.99 mm granules.
5. **Collector:** It will collect all the tea particles in it. The collector size is variable, and the capacity of what is collected will be based on the size of the tea vending machine.
6. **Flavors:** We have three to four flavors with us. The flavors come from natural ingredients; the small grinding/crushing units act individually to add the individual flavors.
7. **Tea Vending Machine:** It will work same as the existing tea vending machine. This machine will collect tea powder from the fresh leaves and will function according to the existing setup. Figure 3.3 provides an overview of the grinding and sieve units. Grinding unit is portable and customizable and can be operated either as a separate one or can be embedded with the existing tea vending machine.

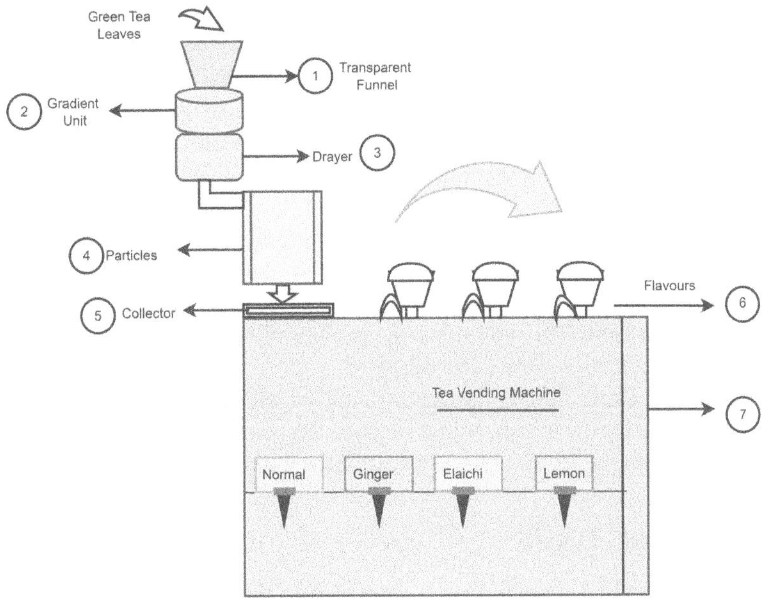

FIGURE 3.2 Design of the tea vendor.

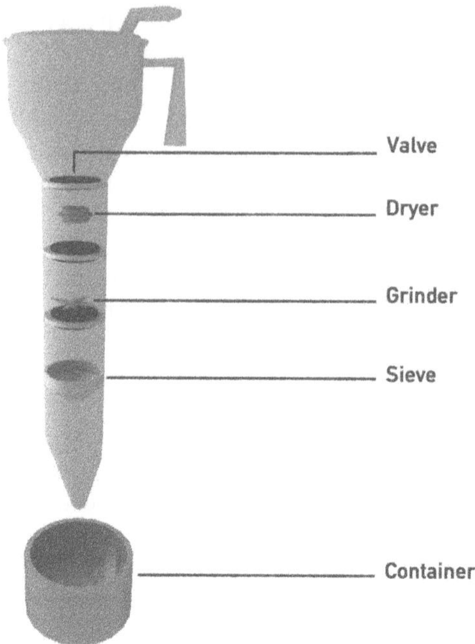

FIGURE 3.3 Grinding and mixing unit.

3.9 COMPLETE SYSTEM

Figure 3.4 depicts the wired diagram of the proposed model. There are two types of containers: one for processing normal tea leaves and other for processing the other leaves, such as jamun, papaya, mango, lemon, guava, tulsi, mint, lemon grass, and hibiscus for various health benefits. This machine can make tea in various combinations, such as normal tea, naturally flavored tea, tea for health with the help of other natural leaves, tea with sugar/without sugar, and tea with milk/without milk.

1. **Collection Unit:** This unit collects the required ingredients to make tea according to the specifications.
2. **Grinding Unit:** This unit grinds the leaves ingredients as per the required sizes.
3. **Distribution Unit:** This unit takes the responsibility of dispersing the powder/sugar/milk ingredients and water.
4. **Processing Unit:** This is the central unit, which has a processor/memory and controls for the actuators that function the complete system. This unit is attached with the IoT Section to transmit the numbers to the cloud servers.

3.10 WORKING UNITS

As shown in Figure 3.4 various flavors are added instantly in the vending machine as labeled 1-3 in the figure. The natural flavors from ingredients are processed instantly thorough these unit. Label 4-7 are the feedback system used in the machine.

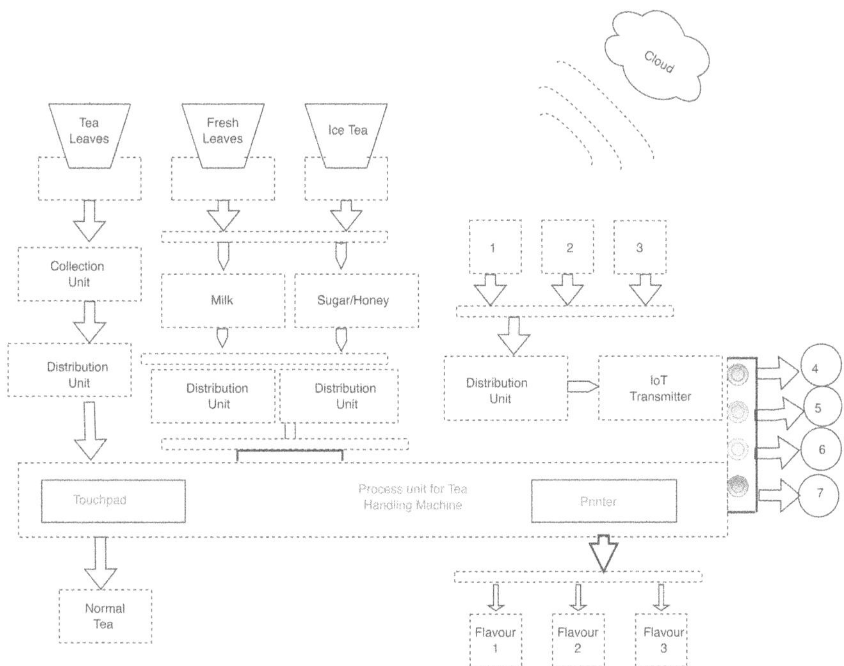

FIGURE 3.4 Complete design of tea vending machine.

The feedback system consists of a set of feedback buttons, which represent scales such as Excellent, Very Good, Good, and Poor.

1. **IoT Transmitter:** This transmitter sends the processed data to the cloud; the framework of data packet is i) Name: Ginger Tea; ii) Quantity: 2; iii) Date Time: 23/11/2019 11.04 A.M.; iv) Place: Chitkara University, Punjab, India; v) Feedback: Excellent. This data is sent to the cloud for analyzing the patterns.
2. **Cloud Server:** The cloud server will receive the data from this tea vending machine, which will be of huge amount since tea vending machines are a source of data generation, and will use for predicting future patterns, such as people's interests, the movement pattern of various kinds of teas, business predictions, advertisement effects, and other types of predictions. Predictions can be made with the available data sets and feedback can be analyzed intensively to improve the overall quality.
3. **Printing Unit:** This unit will give the health benefits of having various kinds of teas, and a small prediction of future health benefits when people continue using the same tea for a week, a month, or a year. The health benefits and good effects of various leaves, such as a reduction in glucose level, blood pressure control, weight loss, and others will attract more customers when things are printed and given to them on the spot.

3.11 MACHINE COMPONENTS

3.11.1 Servo Motor

The vending machine uses a servo motor as an electronic component. The servo motor consists of a rotary and linear actuator to control velocity, acceleration, and angular position. For a particular position, feedback sensor is connected with the motor and controller, but a specific design module is required to use a servo motor.

3.11.2 Arduino Uno Microcontroller

AT mega328P is the microcontroller-based board Arduino Uno. It uses 14 input and output digital pins. Out of 14 pins, 6 are used as the analog inputs, 2 are used as an ICSP header and a reset button, whereas the other 6 are used as Pulse Width Modulation (PWM) outputs. Apart from that, the reset button, power jack, In-Circuit Serial Programming (ICSP) header, Universal Serial Bus (USB) connection, and 16 MHz quartz crystal are connected with the computer, either via an AC to DC adapter or a battery just to get started [46].

3.11.3 Water Heating Unit

It consists of the following units:

1. Water heating element
2. Cut-off thermostat
3. Solenoid valve

The resistive unit is used for the water heating element. The heating coil is heated when the electric current is passed through it, and with the coil, water presents in the bottle gets heated. The thermostat in the heating unit ensures that the temperature does not increase above the threshold value. The main objective of the solenoid valve is to ensure that the required quantity of water reaches the cup, and valve closes and opens on time.

3.11.4 Container

Food-grade plastic steel and food-grade plastic [47] are used to make containers [48]. Material in the cup contain a powder mix that arranged and controlled by the container. The quantity of the cup is controlled by using the crank mechanism. The container contains the prefix powder, and its quantity is 154 ml.

3.11.5 Appliance Body

Mild steel is used to make the appliance body. It also provides the structure and support to all the components [49]. Material selection is described in Table 3.1.

TABLE 3.1
Description of Material Selection

Sr. No.	Name of Components	Material Used
1.	Container	Food-grade steel or plastic
2.	Servo motor	Standard component
3.	Connecting rod	Mild steel or plastic
4.	Crank	Mild steel or plastic
5.	Body	Mild steel
6.	Heating coil	Nichrome

Source: Adapted from [50].

3.12 INNOVATION AND SPECIFIC FUNCTIONS PERFORMED

Table 3.2 lists various innovation and functions.

3.13 FUTURE ENHANCEMENT

There are many automatic tea vending machines, but they use tea powder as the ingredient. In general, tea powders have a lot of artificial flavors and preservatives and are not good for health. Further, consumers do not have any idea of the freshness of tea powders because they feel fresh after taking the tea. Coffee vending machines are available today that prepare coffee from the beans, but they do not require any drier since coffee beans are available as dried beans. In case of tea, there is powder in the unit which is refilled when gets consumed completely. In future work, there should be a mechanism that will show the container is about to become empty and it should be filled. In the proposed machine, our unit has

TABLE 3.2
Description of Functions

Sr. No.	Innovation and Specific Function Performed
1.	Freshness
2.	No loss of natural flavor
3.	No loss of natural flavor
4.	Replace of old powders
5.	Fast
6.	Portable
7.	Customisable
8.	Can be fixed with the existing tea vending machines
9.	Natural flavors without any preservatives
10.	Healthy
11.	Global market
12.	Corporate offices/institutions/hospitals/markets/shopping malls/airports/railway stations

been modified, and in this, we have added a drier/heater unit that will complete the process of preparing fresh tea. It will also provide a health monitoring assistant that can further work as a recommendation system and can give users extra features that currently do not exist in the existing system.

3.14 CONCLUSION

Our proposed model is completely IoT-enabled, and we take advantage of cloud and AI support to improve the efficiency of the overall process. The system can collect user input such as Age: 38, Sex: Male/Female, Weight: 80 kg, Smoking: Yes/No, Drinking: Yes/No, and using AI, it will suggest the type of tea the user can take for better health benefits. This predictive model and suggestive tea vending system is very unique and not found in any existing systems. Additionally, some extra functionality can be added as a future scope so that consumers can gain additional benefits from the proposed machine.

REFERENCES

1. Data Transfer Standard EVA DTS 6.1.1, Dec 2010, European Vending Association AISBL, 44 Rue Van Eyck – 1000 Brussels Alharbi.
2. Multi-Drop Bus/Internal Communication Protocol Version 4.2, February, 2011, National Automatic Merchandising Association, 20 N. Wacker Drive, Suite 3500 Chicago, Illinois 60606-3120 USA.
3. T. Yokouchi (2010). Today and Tomorrow of Vending Machine and Its Services in Japan. Proceedings of IEEE, Service Systems and Service Management (ICSSSM), 7th International Conference, pp. 1–5, doi: 10.1109/ICSSSM.2010.5530240
4. Y. Park & S. Yoon (2011). A Comparison Study of Stock-Out Policies in Vending Machine Systems. Proceedings of IEEE, Engineering and Industries (ICEI), International Conference, pp. 1–4.
5. T. C. Poon, K. L. Choy, C. K. Cheng, & S. I. Lao (2010). A Realtime Replenishment System for Vending Machine Industry. Industrial Informatics (INDIN). 8th IEEE International Conference, pp. 209–213, July 10, doi: 10.1109/INDIN.2010.5549432
6. L. Atzori, A. Iera, & G. Morabito (2010). The Internet of Things: A Survey. Computer Networks, 54, pp. 2787–2805.
7. Z. Wen & Z. X. Long (2010). Design and Implementation of Automatic Vending Machine Based on the Short Massage Payment. Proceedings IEEE, Wireless Communications Networking and Mobile Computing (WiCOM). 6th International Conference, pp. 1–4, doi: 10.1109/WICOM.2010.5600192
8. M. Jovanovic & M. Organero (2011). Analysis of the latest trends in mobile commerce using the NFC technology. Journal of Selected Areas in Telecommunications (JSAT), 1–12.
9. C. Wenshan, H. Yanqun, & L. Minyang (2015). Influential Factors of Vending Machine Interface to Enhance the Interaction Performance. 8th International Conference on Intelligent Computation Technology and Automation (ICICTA), IEEE.
10. V. Vaid (2014). Comparison of Different Attributes in Modeling a FSM Based Vending Machine in 2 Different Styles. International Conference on Embedded Systems (ICES), IEEE.
11. K. Kwangsoo (2014). Smart Coffee Vending Machine Using Sensor and Actuator Networks. IEEE International Conference on Consumer Electronics (ICCE), IEEE.

12. PSG College of Technology (2007), Design Data Book, Coimbatore.
13. M. Zhou, Q. Zhang, & Z. Chen (2006). What Can Be Done to Automate Conceptual Simulation Modelling? Proceedings of the 2006 Winter Simulation Conference, pp. 809–814. International Journal of VLSI design & Communication Systems (VLSICS), 3(2), April 2012, p. 27.
14. B. Roy & B. Mukherjee (2010). Design of Coffee Vending Machine Using Single Electron Devices, Proceedings of 2010 International Symposium on Electronic System Design, pp. 38–43.
15. C. J. Clement Singh, K. Senthil Kumar, Jayanto Gope, Suman Basu, & Subir Kumar Sarkar (2007). Single Electron Device based Automatic Tea Vending Machine. Proceedings of International Conference on Information and Communication Technology in Electrical Sciences (ICTES 2007), pp. 891–896.
16. P. Smith (1997). Automatic Hot-Food Vending Machine. Trends in Food Science & Technology, 81, October 1997, p. 349.
17. M. Zhou, Y. J. Son, & Z. Chen (2004). Knowledge Representation for Conceptual Simulation Modeling. Proceedings of the 2004 Winter Simulation Conference, pp. 450–458.
18. J. Komer (2004). “Digital Logic and State Machine Design,” 2nd ed., Oxford.
19. M. Ali Qureshi, A. Aziz, & H. Faiz Rasool (2011). Design and Implementation of Automatic Ticket System Using Verilog HDL. Proceedings of International Conference on Information Technology, pp. 707–712.
20. S. Kilts (2004). “Advanced FPGA Design: Architecture, Implementation, and optimization,” Wiley-IEEE Press.
21. S. B. Z. Azami & M. Tanabian (2004). Automatic Mobile Payment on a Non-Connected Vending Machine. Proceedings of Canadian Conference on Electrical and Computer Engineering, pp. 731–734.
22. A. Ayman (2020). Vending Machine for Smart Gifting Under-Privileged People. International Journal of Scientific Development and Research (IJSDR), 5(1), 8–14.
23. L. P. Aljadir, W. M. Biggs, & J. A. Misko (1981). Consumption of Foods from Vending Machines at the University of Delaware. Journal of American College Health Association, 30(3), pp. 149–150.
24. I. Baranovski, S. Stankovski, G. Ostojić, D. Oros, & S. Horvat (2018). Software Support for Self-Service Automated Systems”. 17th International Symposium INFOTEH-JAHORINA (INFOTEH), pp. 1–4. IEEE.
25. R. Kondo, I. Harashima, & D. Sunouchi (1989). Automatic Coffee Vending Machine Being Able to Serve a Straight Coffee and a Blended Coffee Selectively, US Patent 4,815,633.
26. M. A. Matthews & T. M. Horacek (2015). Vending Machine Assessment Methodology. A Systematic Review. Appetite, 90, pp. 176–186.
27. S. A. New & M. B. E. Livingstone (2003). An Investigation of the Association between Vending Machine Confectionery Purchase Frequency by Schoolchildren in the UK and Other Dietary and Lifestyle Factors. Public Health Nutrition, 6(5), pp. 497–504.
28. N. T. Nguyen, X. T. Nguyen, J. Lane, & P. Wang (2011). Relationship between Obesity and Diabetes in a US Adult Population: Findings from the National Health and Nutrition Examination Survey, 1999–2006. Obesity Surgery, 21(3), pp. 351–355.
29. C. Ogden, M. Carroll, B. K. Kit, & K. M. Flegal (2014). Prevalence of Childhood and Adult Obesity in the United States”, 2011–2012. Jama, 311(8), pp. 806–814.
30. M. A. Qureshi, A. Aziz, H. Rasool, M. Ibrahim, U. Ghani, & H. Abbas (2011). Design and Implementation of Vending Machine Using Verilog HDL. 2nd International Conference on Networking and Information Technology, IPCSIT, volume 17.

31. S. Saydah, K. M. Bullard, Y. Cheng, M. K. Ali, E. W. Gregg, L. Geiss, & G. Imperatore (2014). Trends in Cardiovascular Disease Risk Factors by Obesity Level in Adults in the United States. NHANES 1999–2010, Obesity, 22(8), pp. 1888–1895.
32. S. Stankovski, G. Ostojić, L. Tarjan, M̌. Stanojević, & M. Babić (2019). Challenges of IoT Payments in Smart Services. Annals of DAAAM & Proceedings, 30.
33. S. Tegeltija, B. Tejić, I. Šenk, L. Tarjan, & G. Ostojic (2020). Universal IoT Vending Machine Management Platform. 19th International Symposium INFOTEH-JAHORINA (INFOTEH), pp. 1–5. IEEE.
34. Branislav Tejić, Sran Tegeltija, Sabolč Horvat, Miroslav Nićin, Milos Štanojević, & Mladen Babic (2019). Payment Methods in Vending Machines. Journal of Mechatronics, Automation and Identification Technology, 4(3), pp. 20–25.
35. NAMA Vision/Industry Definitions, in: www.vending.org/nama_vision/index.php?page=definitions, August 8, 2005, CET 16:47.
36. EVA, European Vending Association: www.eva.be, August 3, 2005, CET 11:13.
37. ÖVV, Österreichische Verkaufsautomaten Vereinigung: www.ovv.at, June 16, 2005, CET 14:07.
38. G. Xie, K. Ota, M. Dong, F. Pan, & A. Liu (2017). Energy-Efficient Routing for Mobile Data Collectors in Wireless Sensor Networks with Obstacles. Peer-to-Peer Networking and Applications, 10(3), pp. 472–483.
39. D. Vasisht, Z. Kapetanovic, J. Won, X. Jin, R. Chandra, S. Sinha, &, S. Stratman (2017). FarmBeats: An IoT Platform for Data-Driven Agriculture. 14th {USENIX} Symposium on Networked Systems Design and Implementation ({NSDI} 17), pp. 515–529.
40. A. Miqdad, K. Kadir, & S. F. Ahmed (2017). Development of Data Acquisition System for Temperature and Humidity Monitoring Scheme. 2017 IEEE 4th International Conference on Smart Instrumentation, Measurement and Application (ICSIMA), pp. 1–4. IEEE.
41. T. AlSkaif, B. Bellalta, M. G. Zapata, & J. M. B. Ordinas (2017). Energy Efficiency of MAC Protocols in Low Data Rate Wireless Multimedia Sensor Networks: A Comparative Study. Ad Hoc Networks, 56, pp. 141–157.
42. K. M. Modieginyane, B. B. Letswamotse, R. Malekian, & A. M. Abu-Mahfouz (2018). Software Defined Wireless Sensor Networks Application Opportunities for Efficient Network Management: A Survey. Computers & Electrical Engineering, 66, pp. 274–287.
43. S. Sethi & R. K. Sahoo (2019). Design of WSN in Real Time Application of Health Monitoring System. In "Virtual and Mobile Healthcare: Breakthroughs in Research and Practice," pp. 643–658. IGI Global.
44. H. Kaur, M. Atif, & R. Chauhan (2020). An Internet of Healthcare Things (IoHT)-Based Healthcare Monitoring System. In "Advances in Intelligent Computing and Communication," pp. 475–482. Springer.
45. V. Kukreja & P. Dhiman (2020, September). A Deep Neural Network Based Disease Detection Scheme for Citrus Fruits. International Conference on Smart Electronics and Communication (ICOSEC), pp. 97–101. IEEE.
46. BDV, Bundesverband der Deutschen Vending Automaten-Wirtschaft e.V.: www.bdvonline.de, June 16, 2005, CET 13:54.
47. H. Diller, (2001). Vahlens Großes Marketinglexikon, Verlag C. H. Beck, Munich 2001, p. 1830.
48. NAMA Vision/Industry Definitions, in: www.vending.org, August 3, 2005, CET 12:19.
49. N. Monssen (1999). Vending – Ein Markt mit Zukunft. BDV (Bundesverband Deutscher Verpflegungs- und Vending-Unternehmen e. V.) (Hrsg.), Köln.
50. H. M. Jungbluth (2002). High-Tech Contra Anonymität. In: gv-praxis Nr. 9, September 4, 2002, p. 64 (translated by authors).

4 Recent Trends in OCR Systems: A Review

Aditi Moudgil, Saravjeet Singh, and Vinay Gautam
Chitkara University Institute of Engineering & Technology, Chitkara University
Punjab, India

CONTENTS

4.1 INTRODUCTION

Optical character recognition (OCR) is one of the most emergent applications of pattern recognition. It is a fully grown, advancing, energizing, and helpful field and is going about as an establishment stone for emerging advanced applications, such as those in the medical field, speech recognition, biometric recognition, etc. (Babu and Soumya, 2019). OCR is a technique that is used to convert scanned images like handwritten text or PDF files into an editable format. In a nation like India, a plenitude of data and invaluable information is available as original copies, old writings, books, manuscripts, family data, and so on. They are customarily accessible either in printed or handwritten formats, and such literature becomes inefficient when information is searched among thousands of pages. Today's era is an advanced era that accepts all the documents and

DOI: 10.1201/9781003143468-4

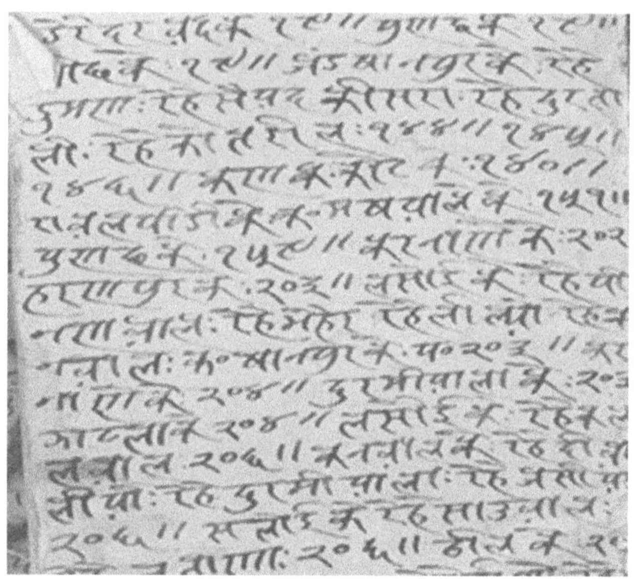

FIGURE 4.1 Devanagari sample manuscript.

available information in a digitized form. Thus, the process of digitizing such literature came to the rescue. The conversion of digitized information into a textual form becomes a challenging task, but it is mandatory to make information readable by any machine processing a million pages per second. When it comes to analyzing manuscripts that might not even be in readable form and the text is deteriorated, it becomes even more challenging to get that precious text and make it available. The old documents that sometimes look blurry, have been inked over, or having missing characters, making them difficult to read, often use character recognition, which is an art of segregating characters from the complete word. The next section shows the application areas of OCR. A sample of Devanagari script is shown in Figure 4.1.

4.1.1 Application Areas of OCR

- Digitizing libraries
- Legal billing documentation
- Data extraction
- Language translation
- Postal address recognition

4.1.2 Working of OCR System

Figure 4.2 represents a typical model for the OCR system. The main phases of OCR include image acquisition (collection of images and their digitization), pre-processing (enhancement, noise removal, slant, and skew correction),

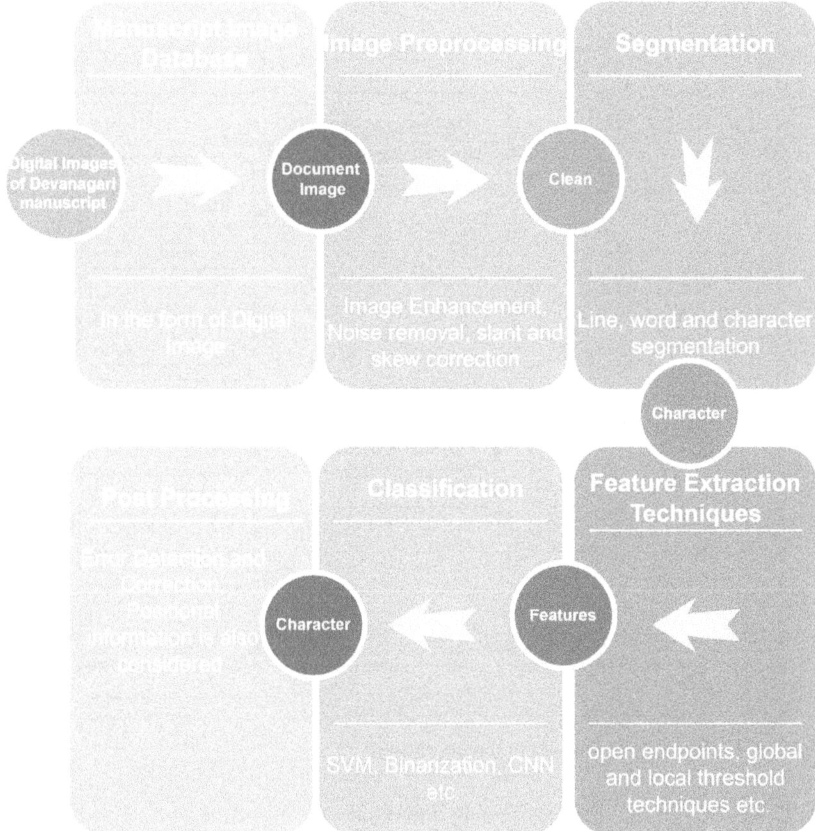

FIGURE 4.2 Working model of a typical OCR.

segmentation (can be classified as page segmentation, line segmentation, word segmentation, and character segmentation), feature extraction (multiple techniques have been implemented, such as zoning, aspect ratio, directional features, projection features, etc.), classification (various classification techniques, such as Multi Layer Perceptron (MLP), Convolutional Neural Network (CNN), Support Vector Machine (SVM), Artificial Neural Network (ANN), Binary tree, K-nearest neighbor (KNN), etc.), and post-processing. Figure 4.3 shows the detailed process of document analysis.

4.1.3 Various Issues Encountered in OCR

1. Noisy and degraded documents make pre-processing a complex task.
2. Touching and overlapping characters make segmentation difficult.
3. Thick and uneven characters make segmentation and recognition very difficult.
4. Faded documents make recognition very complex.

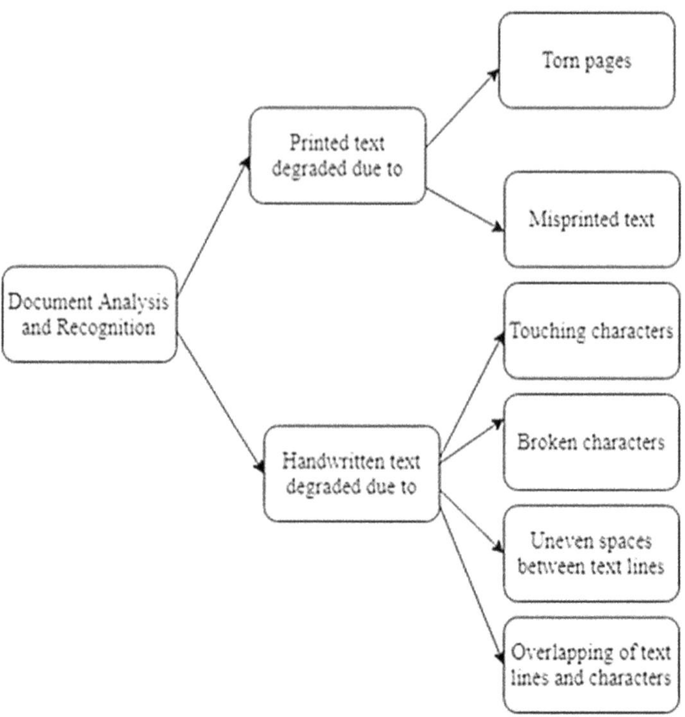

FIGURE 4.3 Document analysis and recognition.

5. Collection of the database, being a manual task, is tedious.
6. Feature extraction and classification techniques must be chosen with care and after thorough experimentation.

4.1.4 Potential Areas for Research

Manuscripts are invaluable assets of a community that contain valuable information. The data preserved in these manuscripts is a treasure that can be extracted to be implemented in our daily lives. The data present is degraded due to certain reasons such as torn pages, overwriting, ink stains, missing headlines, etc.

- To make the data readable, different mechanisms can be employed that motivate the research idea of deploying a system to recognize ancient characters.
- Many of the systems have already been built up to drill down on the ancient characters, thus making it easy to read manuscripts in different languages, such as Devanagari, Sanskrit, Telugu, Malayalam, English, Gurmukhi, Chinese, etc.
- Various feature extraction techniques, such as context information, transfer learning, dataset augmentation, zonal features, shape transformations,

curvelet transformations, wavelet transform coefficients, etc., have been carried out to achieve higher accuracies.

- Also, various classifiers, such as ANN, CNN, feed-forward neural network, SVM, KNN, Maximum Mean Discrepancy (MMD), MLP, and Ada boost, have already been used, but a combination of any of the above-mentioned features and classifiers can be used to optimize the results of character recognition.
- The variability of datasets can also be considered as one of the factors to improve the character recognition rate.

4.1.5 Segmentation

Segmentation, in general terms, means to decompose the image into sub-images so that the background and foreground pixels can be separated. The division stage isolates the original copy picture in different legitimate parts. This is the first and foremost step to develop an OCR, and if it is not done correctly, all other levels of accuracy are affected. So, this step should be carried out most carefully followed by post processing (Kumar et al., 2015). Segmentation is further divided into multiple phases:

1. Page segmentation
2. Line segmentation
3. Word segmentation
4. Character segmentation

4.1.5.1 Page Segmentation

Kumar et al. (2015) in 2013 defined content zones and non-content zones in a report picture. Page division is used to segment content zones from non-content zones. Saini et al. (2019) proposed to have a competition based on the Chinese Historical Reading Challenge in which coders were told to finally make a system that can read Chinese family records. For this evaluation, a dataset consisting of more than 10,000 pages was provided by family search. The extent of this challenge was to fragment the page into multiple classes by allotting an alternate pixel value for each class.

4.1.5.2 Line Segmentation

Line segmentation means splitting the manuscripts into lines in which a top-down approach uses the concept of larger manuscripts that are further split into smaller pieces, which makes it easy to segment lines, whereas the bottom-up approach uses the opposite concept, where smaller pieces are segmented for lines of text and are gathered to form a larger manuscript, thus providing better results. The hybrid approach is a mixture of top-down and bottom-up approaches, which gives the best results (Narang et al., 2019a).

4.1.5.3 Word Segmentation

Word segmentation is a process of separating words from lines that were segmented via line segmentation. Specifically, a word is a collection of multiple characters.

4.1.5.4 Character Segmentation

Character segmentation is the last step performed by OCR in which each character is segmented followed by the recognition of characters. Section 4.2 shows the literature review for manuscript recognition.

4.2 LITERATURE REVIEW

This section covers the review of different techniques used in different parts of OCR models.

4.2.1 Page Segmentation

Kumar et al. in 2013 defined content zones and non-content zones in a report picture. Saini et al. (2019) proposed to have a competition based on the Chinese Historical Reading Challenge in which coders were told to finally make a system that can read Chinese family records. For this evaluation, a dataset consisting of more than 10,000 pages was provided by family search. The extent of this challenge was to fragment the page into multiple classes by allotting an alternate pixel value for each class.

4.2.2 Line Segmentation

Completely isolated lines are really easy to find, but problems can arise for the following reasons:

- Skewed lines
- Overlapping lines
- Touching components
- Curvilinear lines
- Broken characters
- Lines are degraded due to noise

There are three approaches followed for line segmentation:

1. Top-down approach
2. Bottom-up approach
3. Hybrid approach

Arivazhagan et al. (2007) proposed a new technique with the concept of painting. The document was decomposed into vertical stripes out of which each row was painted with the intensity pixel values. Piece-wise potential separation of lines was done. The partial projection was used to detect the line number and skewness of the lines. Also, horizontal borders were provided to segment the text. The contouring method was carried out in the same direction in which the text was written and also in the opposite direction.

Arivazhagan et al. (2007) proposed a method for skewed lines that are also sometimes overlapping and worked on Arabic documents. The author traversed the lines around the handwritten documents. Zahour et al. (2001) used a partial contour-based method to segment lines from manuscripts. Liwicki et al. (2007) used the concept of dynamic programming for text line detection and separation. This novel approach focused on studying the text strokes for both online and offline handwritten texts that used an evade function and added a forfeiture if the identified path was detected closest to the strokes. Misclassified strokes were reported, and thus, text line detection proved to be a difficult task.

4.2.3 Word Segmentation

As explained earlier, we can separate words from the complete lines using the following two methods.

1. Distance-based method
2. Recognition-based method

The distance-based approach is based on calculating the distance between two words or connected components. Measurement of this distance metric can be done using various units, such as Euclidian distance (Louloudis et al., 2009), convex hull metric (Lelore and Bouchara, 2011), run-length distance, etc. Based on the distance calculation, the metric units are classified as inter-word or intra-word distances. Whereas, in the recognition-based method, boundaries are provided to each word, thus making it easy to recognize.

Angadi and Kodabagi proposed an innovative technique to segment Kannada historical documents into lines following by word and character segmentation. The proposed technique works even for modifiers and overlapping lines. Extraction of the word made the use of k-means algorithm that calculated the threshold value to identify words. Accuracies up to 99% were achieved in all three fields (Angadi and Kodabagi, 2014).

Manmatha and Rothfeder (2005) evidenced that for character segmentation, the first and foremost step should be line segmentation followed by word segmentation. The authors worked on historical manuscripts in which they faced a great challenge in background noise and other aspects. Cleaning the document was done by removing margins. A projection profile algorithm was used to segment the documents into lines, thus producing smaller blobs that may result in bigger partitions called words, and smaller partitions called characters (Manmatha and Rothfeder, 2005). Saha et al. (2010) wrote literature in which the ultimate goal was to segment the words into characters. A Hough-based efficient method was designed for standardized performance. The new technique published was applied to low-quality camera captured images. The accuracy measured for word segmentation was 85.7% (Saha et al., 2010).

4.2.4 Character Segmentation

Character segmentation is the last step performed by OCR in which each character is segmented followed by the recognition of characters.

Sonika Narang in 2019 presented a paper in which the author designed an ancient manuscript recognition system able to recognize the Devanagari script, which is currently in fragile condition. In this paper, the author proposed a system that was based on the following steps: digitization, pre-processing, segmentation, feature extraction, and classification. In the first step, the author converted the hard copy of the manuscripts into electronic form. In the next step of pre-processing, two conditions were applied: using the autocorrect feature of Office Image Viewer, incomplete words were attempted to be completed; secondly, binarization was done that converted the electronic image into binary format. In segmentation, line was segmented to form words, and words were further broken down to form characters that could be later on processed to form a complete recognition system. Vertical and horizontal projection methods were applied. For character recognition, the headline also called Shirorekha was separated and then the overlapping characters were figured out. The threshold value calculated in the previous step was used to compare the aspect ratio to find the overlapping characters (Narang et al., 2019b).

Narang et al. (2019) worked on the dataset consisting of 100 pages written in Devanagari and followed multiple steps to make the ancient data readable with the highest accuracy and reliability. With the pre-segmented set of images, the steps followed for this work include feature extraction followed by the classification method, which is responsible for maximizing the accuracy, specificity, and sensitivity of data. In the last step, the author tried to improve the accuracy by using AdaBoost and bagging methods to improve the recognition results. Implementing these methods separates the high- and low-frequency components, out of which a few low-frequency components are meaningful that are selected as features. Classification is a very significant method that identifies to which class an image will belong (Narang et al., 2019a).

Kumar et al. (2018) presented an OCR to recognize Devanagari manuscripts collected from different libraries and museums. The process worked out in different phases. In the first phase of digitization, the image was converted into a grayscale image followed by a segmentation phase, in which the pages were segmented into lines followed by words and finally characters. For line segmentation, the page was divided into vertical lines and the average height was calculated, which resulted in the identification of under-segmented and over-segmented lines. The next step was word segmentation, which was based on the black color pixel intensity. Character segmentation proved to be the most difficult step as it was done without removing the shirorekha. For this, the author developed a new method called the drop flow method. This step took place in multiple iterations for which a hypothetical drop of water was forced from downward to upward; if the drop of water finds its way to move up, then there was no shirorekha, but on the other hand, if the water drop finds no way up, then there may be a case of touching characters or shirorekha. For the recognition of touching characters, an accuracy of about 96% was obtained (Narang et al., 2019b). M. Buchler emphasized Coptic texts – ancient Egyptian scripts. Two different fonts of Coptic texts were experimented upon (Bohairic and Sahidic). The character sets that have been experimented with were taken from several published manuscripts and the problems faced were the different frequency of characters, punctuations, page breaks, and line indications. The Coptic font pages were divided into testing and training data such that each page was considered as a part of training data. For pre-processing, the ScanTailor method was used to eliminate the borderlines and page cutting. The output of pre-processing was tested by the coptologist, and a transcript was generated. The transcript generated was proofread and the necessary corrections were made. Thirty thousand training steps were carried out to train the system. Testing pages were evaluated based on the training data and an accuracy of greater than 99% was achieved after following the above-mentioned steps (Narang et al., 2019a). Palakollu et al. (2012) presented a new method for the recognition of Devanagari documents. In this method, for line segmentation, an average line-height of 30 pixels (px) was considered due to the variation in the handwriting of different people.

Here, line segmentation was based on the horizontal projection method. The word segmentation process was considered somewhat easier as the minimum 3 px distance was considered in-between two words. In zone segmentation, the actual header line was compared to the expected header line, and subsequently, the header line was straightened and the three segments were separated – upper modifier, header line, and other characters. As a result, an accuracy of 93.6% was figured out while segmenting a line; a subsequent accuracy of 98.6% was calculated while segmenting a word (Palakollu et al., 2012).

Pengcheng et al. (2017) worked on Chinese calligraphy, which is written in multiple styles and is not recognizable easily except for the new learners. This paper proposed a system that automatically examines Chinese calligraphy. Three types of feature extraction techniques have been used in this field, namely, the global feature Global Image Descriptor (GIST), local feature Scale-Invariant Feature Transform (SIFT) descriptor, and scale-invariant feature transform technique. Also, three different deep feature extraction classifiers have been used – Convolutional Neural Network (CNN), Support Vector Machine (SVM), and neural network. These three classifiers are made to implement on pre-existing datasets, the unconstrainted real-world calligraphic character dictionary (CCD) consisting of historical calligraphies, and the standard calligraphic character library (SCL). CCD consists of more than 110,000 character images while SCL consists of 18,770 character images. When these tests were done on two different datasets, an accuracy of 99.78% was achieved on the SCL dataset, and comparatively a lower accuracy of 94.22% was achieved on the CCD dataset (Pengcheng et al., 2017).

Yang et al. (2018) proposed a new method named recognition guided detector (RGD) to recognize dense and tight characters in Chinese historic documents. The architecture of this method divides it into three parts, text line segmentation, proposal generation, and character-level detection. The input image is first segmented into text lines using vertical projection. The recognition guided proposal method) is used for character generation from the segmented text lines. Finally, the précised boundary box around the characters is obtained by using RGD. Irrespective of many approaches, the vertical projection approach has been used for text line segmentation where each input image is divided into vertical text lines to avoid confusion. The experiments were performed on two datasets: the Multiple Tripitaka in Han (MTH) and Tripitaka Koreana in Han (TKH) dataset, respectively. For character-level detection, the authors developed a new approach to RGPN based on multiple layers of CNN.

Mahto et al. (2015) present a new technique for skewed word detection and correction. If the aspect ratio of the word comes out to be less than 0.7%, then the word is considered skewed, and for its correction, different algorithms are followed. The word is rotated clockwise and anticlockwise, and finally, the height is calculated. The point where the minimum height was calculated is noted, and the same is done for word width as well. The slope of the word is used to determine the skewness, and as a result, the word is straightened. Clanuwat et al. in 2018 worked on applying machine learning techniques on three datasets, namely, Kuzushiji (cursive Japanese), Kuzushiji-49, and Kuzushiji-Kanji from classical Japanese literature. Multiple machine learning techniques were applied such as the KNN, CNN, and manifold methods, and the expected outcome was measured, which was lying between 95–99% depending on the dataset experimented upon (Clanuwat et al., 2018).

Nguyen et al., 2017 worked on the query by string approach to recognize keywords in Japanese manuscripts and used the CNN method for feature extraction and accuracy measurements. First, the Japanese documents were converted into images to apply the CNN method further, and the results were compared with the papers using the query by example approach. The paper shows that the query by string method is more accurate when compared to the state-of-the-art methods. The authors tried to read the deformed characters of Japanese manuscripts by applying the procedure in three levels. For level one, CNN in a combination of 1-dimensional Long Short-Term Memory (LSTM) has been implemented in 20 epochs, followed by level two, in which CNN in combination with 2-dimensional Bidirectional LSTM (BLSTM) was applied recurrently to improve the performance. In level three, CNN without pre-segmentation was applied with 2-dimensional BLSTM. Rajithkumar et al., (2015) proposed the character recognition rate of stone inscriptions in Kannada by following the steps given ahead.

Stone inscriptions were captured; a noise removal technique was applied to filter out the noisy images followed by edge detection, and a thickening of edges was done to read the text properly. Mean, standard deviation, and the sum of absolute difference algorithm were applied to 40 datasets consisting of 16 characters to achieve an accuracy of 98.75%. Madhavaraj et al. (2014) developed an OCR that was able to effectively segment the characters of merged old Kanadda manuscripts. Correlation and discrete wavelet transform feature extraction techniques were used and the appropriate dataset was divided into test and train data. The SVM algorithm was implemented, which proved to be 91.2% accurate.

Mohana and Rajithkumar (2014) developed a new technique called the Advance Recognition Algorithm (ARA) to recognize Kannada stone inscriptions from Hoysala and Ganga periods. The ARA method was used to determine the era in which a stone was inscribed by recognizing the font, shape, and size of the stone inscriptions of medieval periods. An accuracy rate of almost 100% was achieved.

Sandhya and Krishnan (2016) discussed the problems for the degradation of old texts and tried to solve the problem of degraded text due to broken characters while rebuilding the broken characters. Old manuscripts in Kannada were collected. Sandhya and Krishnan extracted the zonal features while applying the neural network technique to complete the broken characters. Around 50 features were worked upon. After rebuilding, a recognition accuracy of 98.9% was achieved. The work was carried out on synthetic datasets. Fischer et al. (2010) developed ground truth creation for German manuscripts on some pre-defined German databases. Text areas bounded with polygons were selected, and the difference of Gaussians method was applied for binarization. Based on horizontal inclination, text lines were segmented along with the skewness adjustment. The text was manually corrected using the Java application. The hidden Markov method was used to segment the text lines into words. A sliding window was used for feature extraction. After manual correction, almost 100% accuracy was achieved.

Bertholdo et al. (2009) presented a very easy binarization technique called linearization that worked in four steps: global mean calculation for grayscale images, segmentation of images into horizontal text lines, application of threshold value for each row bearing text, and lastly, the application of Kavallieratou technique. The method followed was unable to produce good results for newspaper clippings (Bertholdo et al., 2009). But it somehow showed good results when it came to the separation of foreground characters and background noise. So, Bertholdo was able to separate the text bearing areas of the page, thus providing the best results. Figure 4.4 shows recent publication trends and the types of dataset. This figure provides a brief information about the recent publication, size of dataset, classification technique used, and accuracy obtained (Avadesh and Goyal, 2018; Azawi et al., 2013; Chamchong et al., 2010; Demilew and Sekeroglu, 2019; Diem and Sablatnig, 2009; Narang et al., 2018, 2021; Nguyen et al., 2017; Sharma et al., 2018; Van Phan et al., 2011).

4.3 DISCUSSION AND ANALYSIS

The graph shown in Figure 4.5 represents the multiple classification techniques used in different time slots, i.e., 2011–2013, 2014–2016, 2017–2019, and 2020–2021.

One can easily interpret from this graph that since 2011, the use of SVM technique has increased and this technique is widely used today. CNN is also an emerging technique these days, and over the past decade, the use of CNN has also been increasing. The studies show that ANN has been used since 2011. MLP and random forest techniques also play a vital role in classifying handwritten text. So, these techniques can even be used today for any kind of classification to be performed on historic manuscripts.

Manuscript type	Dataset Size	Classification Technique	Accuracy
Devanagri	100 Pages	CNN	94%
Geez	22,913 images	CNN	99.39%
Myanmar	60000 Characters	CNN	99%
Sanskrit	11,230 images	CNN	-
Devnagri	5484 Character	SVM + Decision Tree	90.7%
Gujrati	20500 Characters	Naive Bayes	53-68%
Japanes	228334 images	CNN + LSTM	96.80%
German	219,948 Words	RNN + LSTM	93.90%
Nom	12905 Characters	Projection + Vernoi	85%
Palm	-	Binarization	-
Glagolitic	20 Imahes	SVM	70%

FIGURE 4.4 Recent literature in the field of OCR and dataset used.

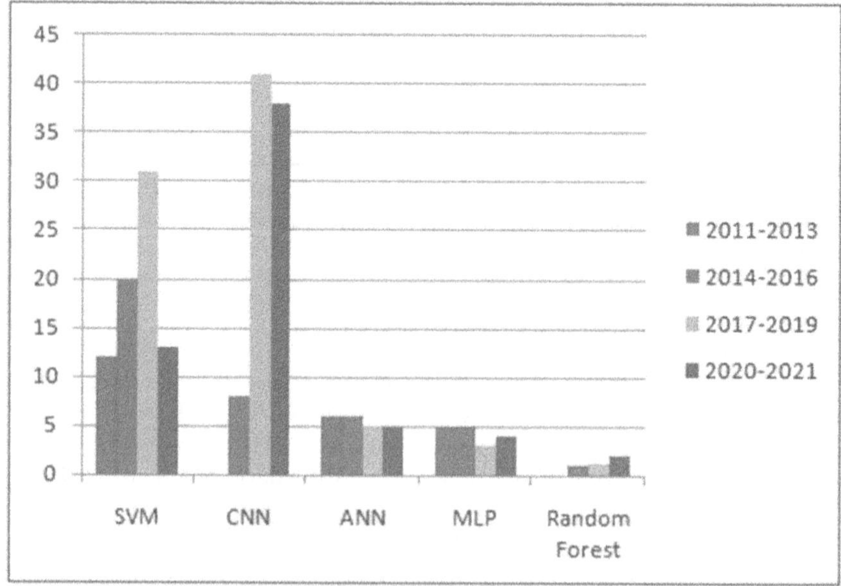

FIGURE 4.5 Year-wise classification technique analysis.

4.4 CONCLUSION

OCR is a process that helps in the identification of characters, either in handwritten or printed text. As shown in the literature, there are many OCR systems that work on the concept of deep learning and use various kinds of feature extraction, segmentation, and classification techniques, along with accuracy measurement techniques, mainly in ancient documents. As reported by many authors, the research in this field is still in its initial stages. The literature presented in this report focuses on various classifications and feature extraction techniques along with the language for which OCR has been designed. The literature focuses on various application areas of the OCR briefly. There is no standard database available for Devanagari manuscripts. Work has already been done on basic characters that can further be extended to modifiers and conjuncts. Lacking behind is the need to design novel OCR for manuscript recognition (Azawi et al., 2013).

REFERENCES

Angadi, S. A. and Kodabagi, M. (2014). A robust segmentation technique for line, word and character extraction from Kannada text in low resolution display board images. *International Journal of Image and Graphics*, 14(01n02):1450003.

Arivazhagan, M., Srinivasan, H., and Srihari, S. (2007). A statistical approach to line segmentation in handwritten documents. In *Document Recognition and Retrieval XIV*, volume 6500, page 65000T. International Society for Optics and Photonics.

Avadesh, M. and Goyal, N. (2018). Optical character recognition for Sanskrit using convolution neural networks. In *2018 13th IAPR International Workshop on Document Analysis Systems (DAS)*, pages 447–452. IEEE.

Azawi, M. A., Afzal, M. Z., and Breuel, T. M. (2013). Normalizing historical orthography for OCR historical documents using LSTM. In *Proceedings of the 2nd International Workshop on Historical Document Imaging and Processing*, pages 80–85.

Babu, N. and Soumya, A. (2019). Character recognition in historical handwritten documents–a survey. In *2019 International Conference on Communication and Signal Processing (ICCSP)*, pages 0299–0304. IEEE.

Bertholdo, F., Valle, E., and de A. Araújo, A. (2009). Layout-aware limiarization for readability enhancement of degraded historical documents. In *Proceedings of the 9th ACM Symposium on Document Engineering*, pages 131–134.

Chamchong, R., Fung, C. C., and Wong, K. W. (2010). Comparing binarisation techniques for the processing of ancient manuscripts. In *Entertainment Computing Symposium*, pages 55–64. Springer.

Clanuwat, T., Bober-Irizar, M., Kitamoto, A., Lamb, A., Yamamoto, K., and Ha, D. (2018). Deep learning for classical Japanese literature. *arXiv preprint, arXiv:1812.01718.*

Demilew, F. A. and Sekeroglu, B. (2019). Ancient geez script recognition using deep learning. *SN Applied Sciences*, 1(11):1–7.

Diem, M. and Sablatnig, R. (2009). Recognition of degraded handwritten characters using local features. In *2009 10th International Conference on Document Analysis and Recognition*, pages 221–225. IEEE.

Fischer, A., Indermühle, E., Bunke, H., Viehhauser, G., and Stolz, M. (2010). Ground truth creation for handwriting recognition in historical documents. In *Proceedings of the 9th IAPR International Workshop on Document Analysis Systems*, pages 3–10.

Kumar, M., Sharma, R., and Kumar, M. G. (2015). *Offline handwritten Gurmukhi script recognition*. PhD thesis.

Kumar, M., Jindal, M. K., Sharma, R. K., and Jindal, S. R., Performance Comparison of Several Feature Selection Techniques for Offline Handwritten Character Recognition. In 2018 *Proceedings of International Conference on Research in Intelligent and Computing in Engineering*, pages 1–6. https://ieeexplore.ieee.org/document/8509076.

Lelore, T. and Bouchara, F. (2011). Super-resolved binarization of text based on the FAIR algorithm. In *International Conference on Document Analysis and Recognition*, pages 839–843.

Liwicki, M., Indermuhle, E., and Bunke, H. (2007). On-line handwritten text line detection using dynamic programming. In *Ninth International Conference on Document Analysis and Recognition (ICDAR 2007)*, volume 1, pages 447–451. IEEE.

Louloudis, G., Gatos, B., Pratikakis, I., and Halatsis, C. (2009). Text line and word segmentation of handwritten documents. *Pattern Recognition*, 42(12):3169–3183.

Madhavaraj, A., Ramakrishnan, A., Kumar, H. S., and Bhat, N. (2014). Improved recognition of aged Kannada documents by effective segmentation of merged characters. In *2014 International Conference on Signal Processing and Communications (SPCOM)*, pages 1–6. IEEE.

Mahto, M. K., Bhatia, K., and Sharma, R. (2015). Combined horizontal and vertical projection feature extraction technique for Gurmukhi handwritten character recognition. In *2015 International Conference on Advances in Computer Engineering and Applications*, pages 59–65. IEEE.

Manmatha, R. and Rothfeder, J. L. (2005). A scale space approach for automatically segmenting words from historical handwritten documents. *IEEE Transactions on Pattern Analysis and Machine Intelligence*, 27(8):1212–1225.

Mohana, H. and Rajithkumar, B. (2014). Era identification and recognition of Ganga and Hoysala phase Kannada stone inscriptions characters using advance recognition algorithm. In *2014 International Conference on Control, Instrumentation, Communication and Computational Technologies (ICCICCT)*, pages 659–665. IEEE.

Narang, S., Jindal, M., and Kumar, M. (2019a). Devanagari ancient documents recognition using statistical feature extraction techniques. *Sādhanā*, 44(6):1–8.

Narang, S. R., Jindal, M. K., and Kumar, M. (2019b). Drop flow method: an iterative algorithm for complete segmentation of Devanagari ancient manuscripts. *Multimedia Tools and Applications*, 78(16):23255–23280.

Narang, S. R., Jindal, M. K., and Sharma, P. (2018). Devanagari ancient character recognition using hog and DCT features. In *2018 Fifth International Conference on Parallel, Distributed and Grid Computing (PDGC)*, pages 215–220. IEEE.

Narang, S. R., Kumar, M., and Jindal, M. (2021). DeepNetDevanagari: a deep learning model for Devanagari ancient character recognition. *Multimedia Tools and Applications*, 80:20671–20686. https://doi.org/10.1007/s11042-021-10775-6.

Nguyen, H. T., Ly, N. T., Nguyen, K. C., Nguyen, C. T., and Nakagawa, M. (2017). Attempts to recognize anomalously deformed kana in Japanese historical documents. In *Proceedings of the 4th International Workshop on Historical Document Imaging and Processing*, pages 31–36.

Palakollu, S., Dhir, R., and Rani, R. (2012). Handwritten Hindi text segmentation techniques for lines and characters. In *Proceedings of the World Congress on Engineering and Computer Science*, volume 1, pages 24–26.

Pengcheng, G., Gang, G., Jiangqin, W., and Baogang, W. (2017). Chinese calligraphic style representation for recognition. *International Journal on Document Analysis and Recognition (IJDAR)*, 20(1):59–68.

Rajithkumar, B., Mohana, H., Uday, J., Bhavana, M., and Anusha, L. (2015). Read and recognition of old Kannada stone inscriptions characters using novel algorithm. In *2015 International Conference on Control, Instrumentation, Communication and Computational Technologies (ICCICCT)*, pages 284–288. IEEE.

Saha, S., Basu, S., Nasipuri, M., and Basu, D. K. (2010). A Hough transform based technique for text segmentation. *arXiv preprint arXiv:1002.4048*.

Saini, R., Dobson, D., Morrey, J., Liwicki, M., and Liwicki, F. S. (2019). ICDAR 2019 historical document reading challenge on large structured Chinese family records. In *2019 International Conference on Document Analysis and Recognition (ICDAR)*, pages 1499–1504. IEEE.

Sandhya, N. and Krishnan, R. (2016). Broken Kannada character recognition—a neural network based approach. In *2016 International Conference on Electrical, Electronics, and Optimization Techniques (ICEEOT)*, pages 2047–2050. IEEE.

Sharma, A. K., Adhyaru, D. M., and Zaveri, T. H. (2018). A novel cross correlation-based approach for handwritten Gujarati character recognition. In *Proceedings of First International Conference on Smart System, Innovations and Computing*, pages 505–513. Springer.

Van Phan, T., Zhu, B., and Nakagawa, M. (2011). Development of nom character segmentation for collecting patterns from historical document pages. In *Proceedings of the 2011 Workshop on Historical Document Imaging and Processing*, pages 133–139.

Yang, H., Jin, L., Huang, W., Yang, Z., Lai, S., and Sun, J. (2018). Dense and tight detection of Chinese characters in historical documents: datasets and a recognition guided detector. *IEEE Access*, 6:30174–30183.

Zahour, A., Taconet, B., Mercy, P., and Ramdane, S. (2001). Arabic hand-written text-line extraction. In *Proceedings of Sixth International Conference on Document Analysis and Recognition*, pages 281–285. IEEE.

5 A Novel Approach for Data Security Using DNA Cryptography with Artificial Bee Colony Algorithm in Cloud Computing

Manisha Rani, Madhavi Popli, and Gagandeep
Department of Computer Science, Punjabi University
Punjab, India

CONTENTS

5.1 INTRODUCTION

Cloud computing is acknowledged as a new prototype that grants on-demand services such as servers, storage, and applications, using a real-time communication network at a minimal cost. It permits vendors to provide reliable and customized information technology (IT) services on a flexible basis using internet via service providers. Previously, computer demanded additional space for servers, network devices, hardware, and other electronic devices for input processing and produced limited output as compared to advanced computers. Nowadays, these valuable electronic devices are replaced by hard drives and reasonable devices. In the '90s, telecommunications

DOI: 10.1201/9781003143468-5

companies proposed point-to-point data circuits to the user; after this, they started offering virtual private networks at a lower cost but with the same quality. Shortly, Salesforce, Amazon and Google brought cloud as a new concept and became key players in the internet marketplace. As many users' applications and data migrate to cloud, the lack of security poses many threats to the cloud platform. Cloud computing follows a layered structure and transfers data through various levels in the cloud environment. Cloud security is considered a significant challenge in the IT industry. Therefore, cloud security mechanisms are implemented at each level. Secure communication is required while transferring data from one point to another. Trust, confidentiality, availability, and integrity are some of the security aspects associated with data.

The main objective is to implement an optimized technique to secure data using DNA cryptography with the Artificial Bee optimization algorithm. The proposed method is classified into two sections: the first section works on the DNA cryptography and the second section works on the Artificial Bee algorithm for key generation. DNA cryptography transforms data into four nucleotides bases: A, G, C, T. The alphanumerical values are encoded and transformed into DNA nucleotides bases. English alphabets and the numerical digits are transformed into random DNA nucleotides sequence, which indicates a regular computer can understand the information encoded by these four sequences as well as the information in a binary sequence (Kalsi, Kaur, and Chang 2018). The Artificial Bee algorithm generates an optimized random key for the encryption and decryption process. The decryption method is a reverse procedure of the encryption process. DNA computing is able to solve many computational problems and provides better performance. This chapter introduces a unique concept of DNA cryptography, whose results are virtually unbreakable, and additionally manage a part of storage issues. A completely random key is generated using the Artificial Bee algorithm that is considered an optimized technique inspired by the foraging behavior of honey bees. This algorithm behaves as a source for a key generation employed for the encryption and decryption methods on the encoded text. The probability is evaluated for the solution, and a new solution is produced from the current results. This process is repeated until the number of cycles approaches its limit.

The rest of this chapter is organized as follows: Section 5.2 presents the existing encryption techniques and approaches applied by multiple authors to protect data from malicious users. Section 5.3 briefly demonstrates the Artificial Bee Colony Optimization algorithm. Section 5.4 presents the proposed algorithm, which starts with DNA computing accompanied by a brief explanation of the encryption and decryption process using the Artificial Bee algorithm for a key generation along with the experimental outcomes in terms of time and file size. Section 5.5 discuses the results of the experiments. Finally, Section 5.6 presents the conclusion.

5.2 RELATED WORK

This section outlines the relevant work based on DNA cryptography and the Artificial Bee Colony (ABC) algorithm. This part is divided into two sections: Section 5.2.1 focuses on techniques based on DNA computing and Section 5.2.2 focuses on the Artificial Bee Colony optimization techniques.

5.2.1 Work Related to DNA Computing Techniques

Modern DNA cryptography has adopted DNA cryptography as a new mechanism for an encryption process. The DNA sequences are transformed and given strength to the existing encryption techniques by adding more elements to the confusion and diffusion process. Different techniques have been applied by multiple authors in order to protect data from intruders, which can be external or internal. In order to secure data from attacks, several approaches are implemented using DNA.

Muthiah and Rajkumar (2014) compared the Genetic algorithm (GA) with the ABC algorithm to minimize the span of the job scheduling process. Overall, the processing time was analyzed and the total processing time in ABC was less than the time taken in the GA. Barkha et al. (2016) proposed the Bi-directional DNA Encryption Algorithm (BDEA) to solve data security problems. DNA bases encode and transform data into American Standard Code for Information Interchange (ASCII) values and later into binary values. Kumar, Iqbal, and Kumar (2016) introduced an algorithm for image encryption. The DNA sequence encoded the Red Green Blue (RGB) image with asymmetric encryption using Elliptic Curve Diffie–Hellman Encryption (ECDHE). Wang and Liu (2017) implemented an image encryption algorithm using chaos and DNA encoding rules. The Piecewise Linear Chaotic Map (PWLCM) generated a key image. Different rows and columns of the plain image were extracted with the logistic map and encoded with the key image using DNA encoding rules. Dokeroglu, Sevinc, and Cosar (2019) proposed the ABC algorithm for solving quadratic problems. In this chapter, employed, onlooker, and scout bees' behavior were modeled by using a distributed memory parallel computation program to solve large and complex problems. Elhadad (2020) proposed a DNA-proxy re-encryption framework to protect cloud data sharing from unauthorized access. In this, three keys were generated: one for the user who wants access, and two others for the owner and the proxy. A user can store encrypted data on the cloud using a key, and if they wish to access data, they can access it via proxy after data re-encryption using a proxy key and decryption using the third key. Kolate and Joshi (2021) proposed a DNA-based security technique as an information carrier. The DNA-based Advanced Encryption Standard (AES) can be used to develop a new data-security system. The proposed framework is aimed to protect records throughout transmission, and is essential whenever a communication or data transmission between sender and recipient has to be confidential. Along with DNA-based AES encryption, a protected DNA-based cryptographic algorithm provides many levels of security. Compression methods can also be used for AES-based DNA cryptography. It can be used to protect confidential material for industrial purposes.

5.2.2 Work Related to Artificial Bee Colony Techniques

Cui et al. (2016) applied the Depth-first search (DFS) framework on the ABC algorithm to speed up the convergence rate problem. The DFS framework allocated more computing resources to the food sources and improves quality. The bee colony is addressed by Intrusion Detection Systems (IDSS) where individuals perform different tasks based on the fitness values. Shah et al. (2018) presents the 3G-ABC algorithm hybrid of Gbest Guided ABC (GGABC) and Global ABC Search (GABCS) for a strong discovery and exploitation process. The implementation is based on the fitness value instead of number of cycles. Lin et al. (2018) designed a novel artificial bee algorithm with local and global search (ABCLGII) to enhance the convergence speed. The local interaction mechanism was applied between employed bees to make searches corporative and directional, and the global interaction mechanism was employed for onlooker bees to exploit good information of some good solutions. Gao et al. (2018) proposed a novel-based ABC algorithm that employed two strategies. One was based on direction learning, and the other was based on elite learning. The direction learning guided searches toward promising areas, whereas elite learning increased the convergence rate without any loss of population diversity. Choong, Wong, and Lim (2019) regulated the search heuristics by applying the modified choice function utilized by employed and onlooker bees. Authors integrated Lin-Kernighan local search strategy to improve the performance of the proposed model. Zhou et al. (2019) proposed a multi-colony ABC algorithm (IDABC) that divided the colonies into three sub-colonies based on the fitness value of the individual as inferior, mid, and superior sub-colonies. In this proposed method, the synergy of the multi-colony was addressed by IDSS where individuals perform different tasks based on the fitness values. Agarwal and Yadav (2019) present a state-of-the-art review of ABC, and it shows recent modifications with an in-depth assessment and analysis of recent common ABC variants. ABC underwent numerous changes to overcome drawbacks and improve its performance in order to become the best for complex optimization problems. Agarwal and Yadav have provided a detailed introduction of ABC and modifications by researchers in detail. Ilango et al. (2019) designed a map/reduce program configured and implemented in a multi-node setting for the ABC algorithm. The main goal of the proposed ABC approach was to reduce execution time and optimization for different dataset sizes. The outcome was evaluated for different fitness and probability parameters determined from the employed and onlooker stages of the ABC algorithm, from which the classification error percentage was further calibrated. The Hadoop environment using map-reduce programming was used to implement the ABC algorithm. The proposed ABC scheme decreased the execution time and classification error for selecting optimal clusters.

Recently, Garg et al. (2020) proposed a new ensemble-based anomaly detection technique for the cloud environment. The authors identified non-linear node behavior among dataset attributes that trigger performance bottlenecks when it comes to detecting malicious performance across various nodes. This paper proposed an ensemble Artificial Bee Colony (En-ABC) based on the anomaly detection scheme for multi-class datasets in a cloud environment based on information. The ABC-based fuzzy clustering technique was used to achieve an optimal clustering based on two objective functions: mean square deviation and Dunn index. Since the amount of data is increasing at such a rapid rate, data clustering using traditional algorithms is becoming increasingly difficult. Taher and Kadhim (2020) proposed a method to improve traditional GA using the ABC algorithm. Earlier, random generation was used to generate an initial population in traditional GA. The authors improved the traditional GA by generating an initial population using ABC algorithm. They used random number generation and traveling salesman problem as a case study for testing the performance of the proposed work. A hybrid of GA and the ABC algorithm provided a good fitness function for the Random Number Generation (RNG) problem. The generated final keys were unique, random, and cryptographically strong. The relative error rate and average tour rate results were better with high convergence rate as compared to traditional algorithm. Shi et al. (2020) proposed an improved artificial honey bee colony for the efficiency of mutual cooperation. The traditional ABC algorithm suffers from slow optimization efficiency and poor performance. Aiming at the defects, the authors implemented an ABC algorithm that initialized honey sources by homogenizing chaotic mapping. The frequency distribution and information entropy were evaluated using a homogenization logistic mapping model. During initial honey point, a homogenization logistic map was used to iteratively produce a suitable global uniform distribution with a suitable approach. Under the condition of limited bee colonies, the algorithm improved convergence performance, optimal solution accuracy, computation efficiency, and reliability. Deng, Xu, and Wu (2021) used ABC algorithm to optimize a block chain investment portfolio. A traditional ABC algorithm solved the single-objective optimization problem. However, the authors improved this algorithm by constructing an external population. The findings showed that the improved ABC algorithm can optimize several features in an investment portfolio at the same time, reduce investor decision-making errors, and enhance the transparency between asset expenses and profits. The ABC algorithm can overcome the issue of portfolio optimization, boost asset securitization security, and improve the balance of investment return and risk to some extent.

5.3 ARTIFICIAL BEE COLONY OPTIMIZATION

Cryptography is an art of achieving security by using mathematical functions to encode messages into non-readable form using different encryption techniques. A cryptographic algorithm works in association with a key to encrypt and decrypt data, and the protocol is known as cryptosystem (Geetha and Akila 2019). Each system must supply some security to protect data from unwanted users. Confidentiality, integrity, authentication, availability, privacy, etc. are some of the goals that can be achieved by cryptography.

Evolutionary and swarm-based optimization techniques have been widely used in recent years. Traditional optimization algorithms are stimulated by natural and social phenomena that are used to solve many complex problems and developed using multiple algorithms, such as the GA, particle swarm optimization, ant colony optimization, artificial bee optimization, and flower pollination algorithm (Saad, Dong, and Karimi 2017). The main objective of the optimization is to achieve the best attainable value using the fitness function and to produce a suitable cryptographic algorithm. The algorithms usually begin with the production of the initial population and new solutions are generated using a solution generated by the previous population. The fitness function evaluates the quality of the solution. This algorithm was proposed by Dervis Karaboga in 2005 to solve optimization problems that were inspired by the foraging behavior of honey bees (Karaboga 2005). Employed, onlooker, and scout bees are three components of the ABC algorithm. In ABC, artificial forager bees search for the food sources. The number of employed bees is equal to the number of onlooker bees around the hive. The employed bees dance on the food sources area and broadcast information to the onlooker bees about the food sources. The onlooker bees elect the food sources found by the employed bees; the food of higher quality will have more chances to be selected by onlooker bees than the one of lower quality. The scout bees got abandoned and converted into employed bees, search for the new food source.

The ABC generates an initial population of randomly distributed swarm solutions denoted by *SN*, where *SN* is swarm size. Let $X_i = \{x_{i,1}, x_{i,2}, x_{i,3}, ... x_{i,D}\}$ be the *i*th solution of swarm and *D* denotes dimensional vector. Each employee bee generates the new solution v_{ij} using Eqn. (5.1).

$$v_{ij} = x_{ij} + rand_{ij}\left(x_{ij} - x_{kj}\right) \tag{5.1}$$

where $k \in \{1,2,\ldots,SN\}$, $j \in \{1,2,\ldots,D\}$, $i \neq j \neq k$, and $rand_{ij} \in [-1,1]$. Each employee bee x_{ij} produces candidate food using Eqn. (5.1). Once the new candidate food is generated, a new fitness value is evaluated. If the fitness value of v_{ij} is better than x_{ij}, then the value of x_{ij} is updated with v_{ij}; otherwise, the old value remains unchanged. After all the employed bees complete the search process, they transmit information to onlooker bees using the waggle dance. The onlooker bees choose the food source depending on the nectar amount with a probability. The richer the source is, the higher the probability of its selection. The probability is a roulette wheel selection mechanism that can be described using Eqn. (5.2).

$$p_i = \frac{fit_i}{\sum_{j=1}^{SN} fit_j} \tag{5.2}$$

where fit_i represents the fitness value of the *i*th solution. If the number of trials reaches its limit and any bee is unable to find a food source, then, in this case, the food source is assumed to be abandoned and scout bees discover new food sources randomly using Eqn. (5.3).

$$P = lower_{bound} + rand(0,1) * \left(upper_{bound} - lower_{bound}\right) \tag{5.3}$$

The ABC algorithm helps to choose the best and optimal solutions from a number of solutions. In this, bees search the neighborhood to find the best solutions. The fitness function can be evaluated for minimization problems using Eqn. (5.4).

$$fit_i = \begin{cases} \dfrac{1}{f_i + 1}, \; f_i \geq 0 \\ 1 + |f_i|, \; f_i < 0 \end{cases} \tag{5.4}$$

The main steps of ABC algorithm are as follows:

Step 1: Generate a random initial population.
Step 2: Evaluate the fitness function of the entire solution and memorize the best solution.
Step 3: Employed bees generate new solutions from the old ones.
Step 4: Evaluate the fitness function in all new solutions and apply the greedy selection method to keep the best solution.
Step 5: Onlooker bees calculate the probability for the solution and generate new solutions from the current solutions depending on the probability and keep the best solution.
Step 6: Abandoned solutions are determined by the scout bees and replaced by randomly generated solutions.
Step 7: Keep the best solution and increment the cycle, cycle = cycle + 1.
Step 8: If the number of cycles reaches its maximum limit, then stop; otherwise go to step 3.

5.4 PROPOSED WORK

The main purpose of the proposed scheme is to provide strong security for cloud data by using a hybrid DNA with the ABC algorithm-based password or secret key, which is generated through several processes. The same key is used for the encryption and decryption process. In the cloud, the proposed scheme will tolerate a variety of attacks. The secret key that enables the device to protect against many security attacks is produced using DNA bases and complementary rules.

In this chapter, the ABC algorithm forms the basis of key generation that will later be used along with the encoded text for the encryption and decryption process. The innovative actions of honey bees have drawn a lot of attention from researchers who aim to invent new approaches. The optimized solution is generated from the randomly generated population. The fitness value of each solution is evaluated, and the best solution is memorized by the bees. The probability for the solution is evaluated and new solutions are generated from the current ones depending on the probability, and thus, bees keep the best solution. This process is repeated until the number of cycles reaches its limit.

Cryptography is one of the prominent solutions for cloud data security. This chapter represents a new hybrid approach for providing security. Due to the complex nature, the theory of DNA has received much interest in the IT era. Unlike typical methods such as 0 and 1, data in the DNA encryption is encrypted and processed using the DNA bases: A (Adenine), C (Cytosine), G (Guanine), and T (Thymine). The random DNA generated key is not secure enough; so to make it secure, this key is embedded with the key generated by the ABC algorithm. The ABC algorithm is the most common evolution optimization algorithm.

5.4.1 DNA Sequence and Encoding

It is a process of determining the nucleotides bases (As, Ts, Cs, and Gs) in a sequence. English alphabets and the numerical digits are transformed into the random DNA nucleotides sequence.

For example, "DNA computing" text is encoded in the random DNA nucleotides sequence using Table 5.1.

5.4.2 Transcription

When the data is encoded into a sequence of DNA nucleotides bases, this encoded text is transcripted into a complementary DNA form (A → U, C→ G, G → C, T→ A).

TABLE 5.1
Random Generated DNA Sequence of Nucleotides Bases

Alphabet	Newly Assigned Value	Numbers	Newly Assigned Value	P	P′
a	ATCG	0	TTCC	d	ACTG
b	CATG	1	CGAT	n	GAGA
c	CTTA	2	TCGA	a	ATCG
d	ACTG	3	CCTT		ATGC
e	CTAG	4	GTTA	c	CTTA
f	TGCA	5	GGTT	o	TAGC
g	GTCA	6	GATC	m	AGAG
h	ACGC	7	ATTG	p	AGGA
i	ACCG	8	GCGG	u	TAGT
j	GCTA	9	GTAC	t	CGCG
k	AGCC	" "	ATGC	i	ACCG
l	TACA			n	GAGA
m	AGAG			g	GTCA
n	GAGA				
o	TAGC				
p	AGGA				
q	ATAG				
r	ATGG				
s	AAGG				
t	CGCG				
u	TAGT				
v	GGAA				
w	TGGT				
x	GACT				
y	TCCT				
z	TGTG				

PP = DNA computing where P′ = ACTGGAGAATCGATGCCTTATAGCAGAGAGGATAGTCGCGAC CGGAG AGTCA

TABLE 5.2
Complementary Strand

Nucleotides Bases	Complementary Strand
A	U
C	G
G	C
U	A

This random sequence is further transformed into a binary sequence. The sequence to convert encoded text into a complementary and binary form is shown in Tables 5.2 and 5.3.

```
P' = UGACCUCUUAGCUACGGAAUAUCGUCUCUCCUAUCAGCGCUGGCCUCUCAGUP' =
111000010111011111001001110001101000001100110110110111011101011
1001101001001100111101001011101110100101l
```

The binary sequence of the encoded sequence is divided into 8-bit block size.

```
P' = 11100001 01110111 11001001 11000110 10000011 00110110
11011101 11010111 00110100 10011001 11101001 01110111 01001011
```

Each byte is right shifted by 2 bits each.

```
P' = 10000111 11011101 00100111 00011011 00001110 11011000
01110111 01011111 11010000 01100110 10100111 11011101 00101101
```

Convert each byte into the decimal equivalent and divide each decimal number with the number generated by the Artificial Bee algorithm. This number will act as the key.

```
D= 135 221 39 27 14 216 119 95 208 102 167 221 45
```

The key generated by the Artificial Bee algorithm is 0.7126563561694389. Operations are performed on the decimal numbers using the key generated using

TABLE 5.3
Random Binary Sequence

Nucleotides Bases	Random Sequence
A	00
C	10
G	01
U	11

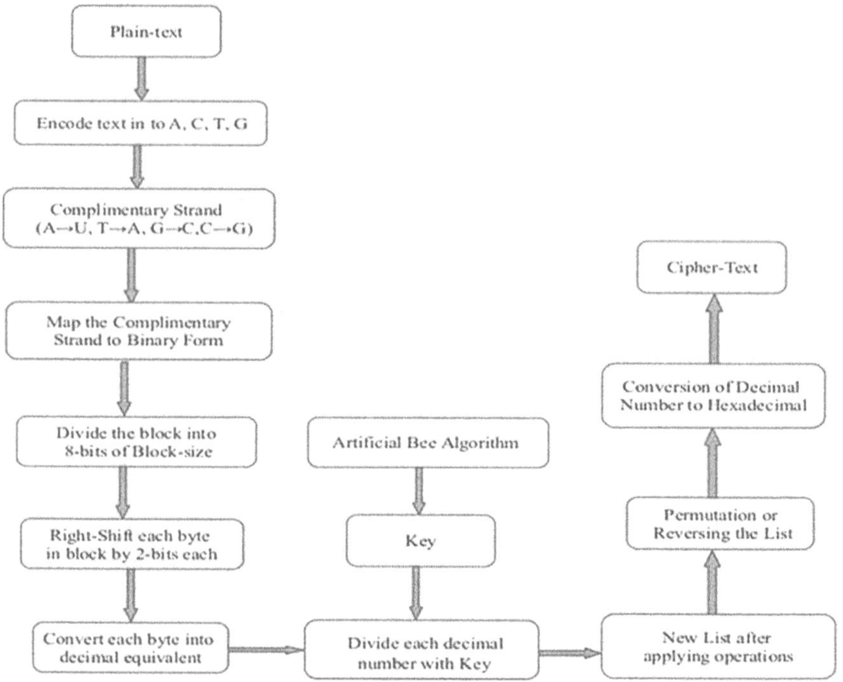

FIGURE 5.1 Encryption process using ABC algorithm.

the ABC algorithm and the result is stored in R. This result is further converted into hexadecimal form. The complete encryption process is shown in Figure 5.1.

This algorithm is implemented using python. Random sets of DNA nucleotides bases (As, Cs, Gs, and Ts) are assigned to the alphabets and numerical digits. This encoded text is transcripted and transformed into a complementary and binary form. The key is generated using the Artificial Bee algorithm and operations are performed on the encoded text using this key. This text is re-encrypted by converting into hexadecimal. The decryption process follows the exact reverse process of the encryption method.

5.5 RESULTS

Many studies are being done to find an optimized solution, in order to meet large amounts of computational storage, and operations that are able to create new methods. DNA computing is considered the best algorithm that does not provide vast storage capacity but increases the complexity, and it is becoming the future for cryptography. This algorithm is based on the DNA computing with the Artificial Bee algorithm. DNA cryptography increases the complexity, where as ABC provides the optimized solution and helps in generating the key to increase the algorithm speed. In this proposed technique, time is considered an influential factor for analyzing purposes during the encryption and decryption process for file sizes greater than 1 KB.

TABLE 5.4
Time Taken for Different File Sizes

File Size (KB)	Encryption Time (ms)	Decryption Time (ms)
3	0.0204	0.0322
6	0.0415	0.0633
10	0.0653	0.0903

This algorithm is designed at the digital level and the time it takes for the encryption and decryption process is shown in Table 5.4. The graph plot is shown in Figure 5.2.

5.6 CONCLUSION

Artificial Bee Colony optimization is as warm-based optimization technique that achieves the best accessible value using the fitness function to generate a suitable cryptographic algorithm. In this chapter, DNA cryptography is used for data encryption along with the Artificial Bee Colony algorithm for key generation. DNA computing is the process of encoding alphabets and numbers into the nucleotides bases (As, Ts, Cs, and Gs) and transcripting and transforming them into a complementary and binary form. DNA computing is able to solve many computational problems and provide better performance. This chapter introduces a unique concept of DNA cryptography, with results that are virtually unbreakable and, additionally, manage a part of storage issues. The key is generated using the Artificial Bee algorithm and operations are performed on the encoded text using this key. This text is further re-encrypted by converting into hexadecimal. The decryption process is the exact reverse process of the encryption process.

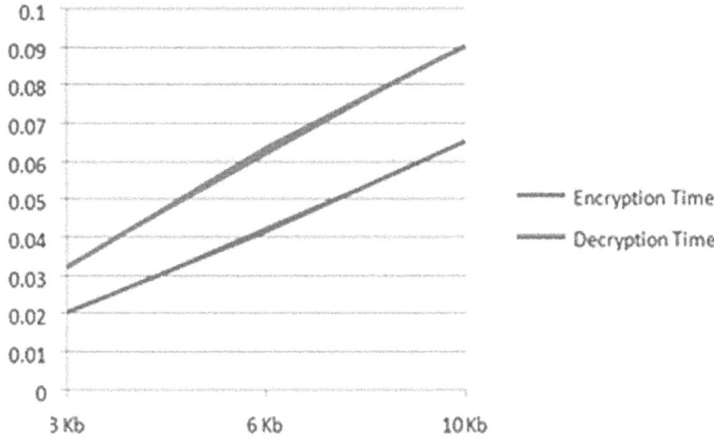

FIGURE 5.2 Encryption and decryption times.

REFERENCES

Agarwal, Shiv Kumar, and Surendra Yadav. 2019. "A Comprehensive Survey on Artificial Bee Colony Algorithm as a Frontier in Swarm Intelligence." *Ambient Communications and Computer Systems*. Springer: 125–34.

Barkha, Prajapati. 2016. "Implementation of DNA Cryptography in Cloud Computing and Using Socket Programming." In *2016 International Conference on Computer Communication and Informatics (ICCCI)*, 1–6. IEEE.

Choong, Shin Siang, Li-Pei Wong, and Chee Peng Lim. 2019. "An Artificial Bee Colony Algorithm with a Modified Choice Function for the Traveling Salesman Problem." *Swarm and Evolutionary Computation* 44: 622–35.

Cui, Laizhong, Genghui Li, Qiuzhen Lin, Zhihua Du, Weifeng Gao, Jianyong Chen, and Nan Lu. 2016. "A Novel Artificial Bee Colony Algorithm with Depth-First Search Framework and Elite-Guided Search Equation." *Information Sciences* 367: 1012–44.

Deng, Yulin, Hongfeng Xu, and Jie Wu. 2021. "Optimization of Blockchain Investment Portfolio under Artificial Bee Colony Algorithm." *Journal of Computational and Applied Mathematics* 385. Elsevier: 113199.

Dokeroglu, Tansel, Ender Sevinc, and Ahmet Cosar. 2019. "Artificial Bee Colony Optimization for the Quadratic Assignment Problem." *Applied Soft Computing* 76. Elsevier: 595–606.

Elhadad, Ahmed. 2020. "Data Sharing Using Proxy Re-Encryption Based on DNA Computing." *Soft Computing* 24 (3). Springer: 2101–8.

Gao, Weifeng, Hailong Sheng, Jue Wang, and Shouyang Wang. 2018. "Artificial Bee Colony Algorithm Based on Novel Mechanism for Fuzzy Portfolio Selection." *IEEE Transactions on Fuzzy Systems* 27 (5). IEEE: 966–78.

Garg, Sahil, Kuljeet Kaur, Shalini Batra, Gagangeet Singh Aujla, Graham Morgan, Neeraj Kumar, Albert Y Zomaya, and Rajiv Ranjan. 2020. "En-ABC: An Ensemble Artificial Bee Colony Based Anomaly Detection Scheme for Cloud Environment." *Journal of Parallel and Distributed Computing* 135. Elsevier: 219–33.

Geetha, M, and K Akila. 2019. "Survey: Cryptography Optimization Algorithms." *International Journal of Emerging Technology and Innovative Engineering* 5 (1): 123–135.

Ilango, S Sudhakar, S Vimal, M Kaliappan, and P Subbulakshmi. 2019. "Optimization Using Artificial Bee Colony Based Clustering Approach for Big Data." *Cluster Computing* 22 (5). Springer: 12169–77.

Kalsi, Shruti, Harleen Kaur, and Victor Chang. 2018. "DNA Cryptography and Deep Learning Using Genetic Algorithm with NW Algorithm for Key Generation." *Journal of Medical Systems* 42 (1). Springer: 1–12.

Karaboga, Dervis. 2005. "An Idea Based on Honey Bee Swarm for Numerical Optimization." 200.

Kolate, Varsha, and R B Joshi. 2021. "An Information Security Using DNA Cryptography along with AES Algorithm." *Turkish Journal of Computer and Mathematics Education* 12 (1S): 183–92.

Kumar, Manish, Akhlad Iqbal, and Pranjal Kumar. 2016. "A New RGB Image Encryption Algorithm Based on DNA Encoding and Elliptic Curve Diffie–Hellman Cryptography." *Signal Processing* 125. Elsevier: 187–202.

Lin, Qiuzhen, Miaomiao Zhu, Genghui Li, Wenjun Wang, Laizhong Cui, Jianyong Chen, and Jian Lu. 2018. "A Novel Artificial Bee Colony Algorithm with Local and Global Information Interaction." *Applied Soft Computing* 62. Elsevier: 702–35.

Muthiah, A, and R Rajkumar. 2014. "A Comparison of Artificial Bee Colony Algorithm and Genetic Algorithm to Minimize the Makespan for Job Shop Scheduling." *Procedia Engineering* 97. Elsevier: 1745–54.

Saad, Abdulbaset El Hadi, Zuomin Dong, and Meysam Karimi. 2017. "A Comparative Study on Recently-Introduced Nature-Based Global Optimization Methods in Complex Mechanical System Design." *Algorithms* 10 (4). Multidisciplinary Digital Publishing Institute: 120.

Shah, Habib, Nasser Tairan, Harish Garg, and Rozaida Ghazali. 2018. "Global Gbest Guided-Artificial Bee Colony Algorithm for Numerical Function Optimization." *Computers* 7 (4). Multidisciplinary Digital Publishing Institute: 69.

Shi, Kexiang, Liyong Bao, Hongwei Ding, Lei Zhao, and Zheng Guan. 2020. "Research on Artificial Bee Colony Algorithm Based on Homogenization Logistic Mapping." In *Journal of Physics: Conference Series* 1624: 42030.

Taher, Ali Abdul Kadhim, and Suhad Malallah Kadhim. 2020. "Improvement of Genetic Algorithm Using Artificial Bee Colony." *Bulletin of Electrical Engineering and Informatics* 9 (5): 2125–33.

Wang, Xingyuan, and Chuanming Liu. 2017. "A Novel and Effective Image Encryption Algorithm Based on Chaos and DNA Encoding." *Multimedia Tools and Applications* 76 (5). Springer: 6229–45.

Zhou, Jiajun, Xifan Yao, Felix T S Chan, Yingzi Lin, Hong Jin, Liang Gao, and Xuping Wang. 2019. "An Individual Dependent Multi-Colony Artificial Bee Colony Algorithm." *Information Sciences* 485. Elsevier: 114–40.

6 Various Techniques for the Consensus Mechanism in Blockchain

Shivani Wadhwa and Gagandeep
Department of Computer Science,
Punjabi University
Punjab, India

CONTENTS

6.1 INTRODUCTION

Today, we are at the verge of technological advancements that will drastically revolutionize our living standards (Kukreja & Dhiman, 2020; Kukreja, Marwaha, Sareen, & Modgil, 2020). In 2008, Satoshi Nakamoto gave origin to bitcoin as a digital currency that basically eliminates the double-spending problem (Nakamoto & Bitcoin, 2008). A blockchain is fundamentally a distributed database of the records of all transactions or electronic events that are executed and shared with the parties involved. Blockchain is a peer-to-peer (P2P) network that is a digital, decentralized, secure, and incorruptible distributed ledger containing information about transactions in financial terms (Mingxiao, Xiaofeng, Zhe, Xiangwei, & Qijun, 2017). It is a linked list of blocks that are cryptographically linked to each other. Figure 6.1 gives an overview of the architecture of blockchain for governance application. Blocks are verified by a consensus mechanism that further needs compute interface for complex computations.

Blockchain technology is becoming more ubiquitous in various applications of information technology. Blockchain is indeed far more than a digital platform for payment. Various companies, such as Ernst & Young, are investing a lot in the

DOI: 10.1201/9781003143468-6

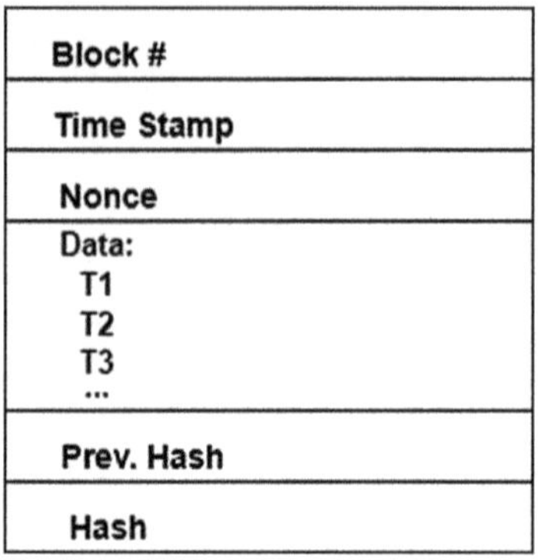

Block #

Time Stamp

Nonce

Data:
T1
T2
T3
...

Prev. Hash

Hash

FIGURE 6.1 Structure of the block.

expansion of privacy tools. It is expected that projects like multiparty computation (MPC), zero-knowledge, etc., will mature enough as they enter the blockchain space. Various IoT deployments will also implement blockchain because of its secure framework. IndiaChain is the ambitious project of Niti Aayog that the Indian government is building to incorporate blockchain into various government projects. Implementing blockchain in different scenarios by working on their use requires huge amount of computations. These computations need an enormous amount of power to solve the consensus mechanism of blockchain. So, there is a need to develop efficient techniques for blockchain computation to increase the acceptance of blockchain in various fields.

Blockchain can provide support to smart devices that are generally low powered and have less capabilities with computation. Huge amounts of computation are required for performing the mining task of block in blockchain. There exists a requirement to perform the computation of blockchain by optimally utilizing resources. So, there arises the need for efficient techniques for the consensus mechanism for blockchain computation. The integration of technologies will ease the deployment of blockchain for any application by providing access to local computing power that supports hashing, consensus mechanism like Proof of Work (PoW), and encrypting algorithms. Although some work has been done to integrate computing technologies with the blockchain, some well-organized mechanisms are required so that computation can be done optimally.

6.2 TYPES OF BLOCKCHAIN

Three types of blockchains exist: public blockchain, private blockchain, and consortium blockchain. All types of blockchains function on a P2P network system with a copy of the shared ledger, which is updated frequently. The variation in how different types of blockchains function is explained as follows:

1. **Public Blockchain:** It is an open-source blockchain in which anyone is allowed to participate as a developer, user, or miner. It does not have any

access restrictions. Public blockchain is decentralized and fully distributed as all transactions are linked in the chain form. Transactions are verified using consensus algorithms. After verification, data modifications are not allowed. Examples of public blockchain are Bitcoin and Ethereum (Lu, 2018).

2. **Private Blockchain:** It is a permissioned blockchain in which consent from the network administrator is required to become part of the network. It is more centralized when compared to public blockchain as only few transactions are given permission to join and become part of the network. As data is private, only invited users can make transactions over it. Examples of private blockchain are Hyperledger Fabric of The Linux Foundation.
3. **Consortium Blockchain:** It is a semi-decentralized blockchain as only preapproved nodes can become part of the network. It offers a higher level of regulation over the network and provides security similar to that of public blockchain. Equal powers are given to all participants of consortium blockchain. Here, multiple enterprises carry out transactions or exchange information on common platforms. Examples of consortium blockchain are Hyperledger, Corda, and Quorum.

6.3 CHARACTERISTICS OF BLOCKCHAIN

Blockchain is gaining worldwide acceptance nowadays because of its use in most of the current technologies and a large variety of applications. It supports various applications through its most favorable characteristics, which are listed below:

1. **Distributed and Decentralized:** Central authority does not possess all the power and information; rather it distributes all the power and information to the various nodes connected with it. Each member inside the network has a duplicate of the precise identical ledger. If a participant's ledger is manipulated by an attacker, it'll be rejected through the bulk of the participants inside the network.
2. **Immutable and Efficient:** Blockchain does not require a third party for appending transactions into the blockchain. In fact, all the nodes present in the network take part in reaching a consensus and then perform the task of verification of block as well. Once block is created, then no change can be made over it.
3. **Anonymous Identity:** This feature of the blockchain makes the user pseudonymous as no one will come to know who is the actual producer of the data, but data will be transparent in nature. It is mostly convenient for providing protection to the producers.
4. **Auditable:** As each of the blockchain transactions are authenticated and documented with a timestamp, blockchain users can very easily identify previous records by accessing any node throughout the distributed network. This feature makes the data traceable and transparent in blockchain.

5. **Secure:** Every user has keys to store the fragment of data in an encrypted form, which provides complete privacy to the data, without involvement of third party. Data is stored in a linear and chronological manner; so each node present in the network stores the hash of its own block along with the hash of the previous block.
6. **Openness:** There are two interpretations of openness in the blockchain. First, it is an open source, i.e., it is available to everyone in the network. Second, any node in the network can participate in appending block into the blockchain.

6.4 APPLICATIONS OF BLOCKCHAIN

Applications that are based on blockchain benefit greatly as they have several unique characteristics when compared with standard databases. Following are a few applications of blockchain:

1. **Smart Healthcare:** The data generated by e-healthcare is huge. Providing security and privacy to such a data is very important, so that patients, doctors, and healthcare practitioners can easily rely on this data. When this set of data is stored in blockchain, it improves the quality of data as well as makes it cost-efficient (De Aguiar, Faiçal, Krishnamachari, & Ueyama, 2020).
2. **Logistic Companies:** As businesses grow in size, it becomes more difficult to manage and track the assets. Asset tracking of logistic companies nowadays is done by using blockchain. The maintenance of transactions and records becomes much easier between the stakeholders (Lao, Li, Hou, Xiao, Guo, & Yang, 2020). Blockchain provides feature of scalability and reliability to logistic companies.
3. **Smart Cities:** Smart cities include heterogeneous networks, a large variety of sensors, and information processing units. It is very important to include blockchain in its architecture, so that data can become secure and users can easily rely on the services provided by smart cities (Sharma & Park, 2018). Embracing blockchain and smart cities together will enhance downstream network processing.
4. **Real State:** The property transaction process can be streamlined easily between the buyer and the seller using blockchain technology. Data tampering is not possible in blockchain-based records. Data recorded is efficient and permanent, which makes the system seamless (Garcia-Teruel, 2020).
5. **Smart Energy Sector:** Microgrids are the storage house of all sources of electric power that manage and increase the overall efficiency (Xue, Teng, Zhang, Li, Wang, & Huang, 2017). Microgrids facilitate the buying and selling of the excess of energy. Blockchain is used in microgrids to maintain the record of transaction between the buyer and the grid as well as between the grid and the customer.
6. **Identity Management:** Digital identities play a very vital role in online transactions for representing the identity of the user. Identity information

for online users needs to be stored securely. Blockchain technology can provide independent, tamper-proof, and secure identity management solutions (Zhu & Badr, 2018).

7. **Insurance:** Nowadays, policies in different insurance companies are automated by using smart contracts. Blockchain allows insurance companies to expand easily by supporting various clients, insurance companies, and many policyholders (Raikwar, Mazumdar, Ruj, Gupta, Chattopadhyay, & Lam, 2018). Blockchain makes the system cost-effective and reduces the complexity process of insurance claims.

6.5 RELATED WORK

A review of related literature provides a survey and discussion of the existing technologies in a given area of study. It provides a concise overview of what has been studied, argued, and established about a topic. Our work focuses on blockchain computation. Literature surveys have been done to gain a basic understanding of blockchain, applications of blockchain, and how to perform blockchain computation on different platforms. The literature review is divided into two subsections:

1. **Based on the Deployment of Blockchain in Various Applications:** Smart devices or mobile devices are producing huge amounts of data. As these devices do not have a great capacity to perform the computations, the data is generally transferred to some other platform where computational tasks can be performed. Nowadays, blockchain is used to provide security features to the data being transferred. Blockchain technology plays a vital role in managing the privacy and security of the IoT devices. Problems that can be addressed by blockchain are identity and access management, data authentication and integrity, authentication, authorization and privacy of users, secure communication, etc. (Khan & Salah, 2018). When blockchain is integrated with IoT devices, it can solve many security problems of IoT. Features of blockchain like auditability, immutability, security, and reliability are favorable for IoT technology and also bring autonomy in IoT devices. The architecture of integrated blockchain and IoT is proposed, which provides a novel IoT platform (Casado-Vara, Chamoso, De la Prieta, Prieto, & Corchado, 2019). An adaptive controller is developed by queuing a theory to achieve the optimal block number to increase the efficiency of the mining process. A new model for increased search speeds via hashmap is also proposed. The data search in the big databases can be improved by using a hashmap stored in the blockchain. Ethereum blockchain has been used to create secure virtual zones or bubbles where smart devices can recognize and trust each other (Hammi, Hammi, Bellot, & Serhrouchni, 2018). Different smart devices are evaluated on the basis of their time and energy consumptions. Optimizing the number of miners in a defined system and the mechanism for selecting miners are left for future work. Hammi et al. (2018) discusses the issues related to integrating IoT with blockchain technology. Advantages of blockchain for large-scale IoT devices are

also stated. Scalability, processing time, legal issues, variation in computing capabilities, etc., are a few challenges that are mentioned in this paper on integrating blockchain technology with IoT (Kumar & Mallick, 2018). This paper identifies the use cases of IoT environment on the basis of the blockchain mechanism. All layers of IoT are explained with security mechanisms and various applications of blockchain (Minoli & Occhiogrosso, 2018).

Suggested work includes identifying the best-suited IoT applications for applying blockchain mechanisms and implementing optimal distributed ledgers for supporting IoT. An approach for managing IoT devices using Ethereum is proposed. Smart contracts using the solidity language are written over the Ethereum platform to manage meter contracts of IoT devices. Investigations show that further studies involving fully scaled multiple IoT devices are also feasible on this platform (Huh, Cho, & Kim, 2017). Challenges that IoT and blockchain together must address are highlighted in the paper of Reyna, Martín, Chen, Soler, & Díaz, 2018. A complete overview of the interaction between the IoT paradigm and blockchain on the basis of existing platforms and applications is also addressed in the paper of Reyna, Martín, Chen, Soler, & Díaz, 2018. It has also been identified that Ethereum is used as the blockchain platform for IoT devices. The integration of IoT and blockchain will greatly increase the use of blockchain, which will ultimately provide scalability, storage capacity, security, and privacy to IoT devices. Mechanisms related to security and offloading in a multi-user mobile edge-cloud computation offloading (MECCO) are proposed in the paper of Nguyen, Pathirana, Ding, & Seneviratne, 2019 for delay-sensitive IoT applications. An access control mechanism for preventing malicious offloading access of cloud resources is used. An optimal offloading policy, which is a Deep-Reinforcement-Learning (DRL)-based offloading scheme for the IoT network, is created. Decisions related to task offloading are considered a joint optimization problem, which is solved efficiently by the Deep Q-learning Network (DQN) algorithm. The framework for a time-sensitive network management service and the related demands of various IoT devices can be enhanced (Nguyen, Pathirana, Ding, & Seneviratne, 2019). A novel architecture is proposed for IoT devices that use edge layer for improving data quality and false data detection using blockchain. This architecture allows decentralized data management via blockchain and computation distribution is done via the edge computing paradigm. This provides optimization to the hybrid end-to-end system (Casado-Vara, Chamoso, De la Prieta, Prieto, & Corchado, 2019).

2. **Based on Different Computing Platforms for Blockchain Computation:** A blockchain-based edge computing framework is designed for offloading computation processes to ensure data integrity. For balancing computations offloading, a Nondominated Sorting Genetic Algorithm (NSGA-III) is used. Simple Additive Weighting (SAW) and Multiple Criteria Decision Making (MCDM) are used as an optimal offloading strategy. Blockchain-based computation offloading (BCO) is proposed for 5G networks (Guo, Hu, Guo, Qiu, & Qi, 2019). It is suggested that the proposed BCO can also

> be adjusted according to the real-world scenario and more users' preferences for Quality of Service (QoS) can be considered. The two-stage stackelberg game model is proposed to maximize the profits of the edge service provider and individual utilities of different miners (Xiong, Feng, Niyato, Wang, Han, 2018).

Nash equilibrium point is derived among the miners, which helps maximize the profit of miners. Uniform and discriminatory pricing schemes are used by implementing backward induction. A platform is proposed for decentralized computation offloading on the basis of blockchain (Seng, Li, Luo, Ji, & Zhang, 2019). A blockchain platform is established for announcing the requests of computation offloading and finding edge servers that conduct offloaded computations. A Galey Shapley (GS)-based user matching algorithm is designed to match the offloading requester's computation task with the edge server. The user matching algorithm is based on execution time and energy consumption. A novel framework for mobile edge computing-enabled wireless blockchain is proposed. Mining tasks that are computation-intensive are offloaded to edge computing nodes, and the cryptographic hash of the blocks are cached in the server of Mobile Edge Computing (MEC) (Liu, Yu, Teng, Leung, & Song, 2018). Stochastic game theory is applied to derive the performance measures of delay, energy consumption, and orphaning probability. Then the algorithm on the basis of alternating direction method of multipliers (ADMM) is applied. It is suggested that other QoS constraints can be considered in wireless blockchain networks. A technique is designed to optimize the cost of mobile equipment (ME) by using joint computation offloading and coin-loaning (Zhang, Hong, Chen, Zheng, & Chen, 2019). An efficient distributive algorithm is designed on the basis of non-cooperative game method. The proposed algorithm quickly achieves the Nash equilibrium (NE) point. For execution, smart contracts are deployed on the Ethereum networks. For further optimization, utilization of stackelberg game has been suggested for MEs, banks, and edge servers at the same time. The BeCome method is designed to achieve load balancing and data integrity of smart devices by decreasing offloading times and energy consumption (Xu, Zhang, Gao, Xue, Qi, & Dou, 2019). The non-dominated sorting genetic algorithm III (NSGA-III) is implemented for possible task offloading schemes. Multiobjective optimization is achieved by implementing SAW and MCDM. An extension of this work can be done by implementing it on real scenarios of the IoT. The auction mechanism is proposed for the edge service provider to assign the edge computing resources that enable mobile blockchain proficiently. The main focus of this work is to maximize the social welfare by providing individual rationality, computational efficiency, and truthfulness (Jiao, Wang, Niyato, & Xiong, 2018). It is also stated that this work can be considered for the various demands of mobile users. A statistical method is proposed to solve the complex mathematical puzzle in PoW (Altman, Reiffers, Menasche, Datar, Dhamal, & Touati, 2019). The mathematical model of expectations and the polynomial matrix factorization method are used. The proposed approach consumes less memory, power, and time by simplifying the system. This model is applicable for all consensus algorithms. Suggested work is to implement the proposed model for hybrid network consisting of irrational multiagents. A neural network architecture based on deep learning is proposed for the allocation of edge resources in mobile blockchain networks (Luong, Xiong, Wang, & Niyato, 2018). Miner bids are given as input to the neural networks and the winner representation and payment of miners represent the output of the neural network. Parameters of the network are optimized by Stochastic Gradient Descent (SGD). This work can also be extended by considering multiple edge computing resource units. An EdgeChain framework is proposed that integrates permissioned blockchain and smart contract with the distributed IoT applications (Pan, Wang, Hester, Alqerm, Liu, & Zhao, 2018). EdgeChain implements a coin system, which is enabled by blockchain to link the resources of the edge pool with the IoT device accounts and the resource usage manners. The implementation of smart contract is done to regulate behavior and enforce policies with IoT devices. This project is ongoing and work is being done for the IoT proxy, intelligent resource provisioning for multiple heterogeneous applications, and the regulation of better IoT device behavior.

6.6 STRUCTURE OF BLOCK

There are various fields that are required for the construction of block. The first block of the blockchain is known as genesis (i.e., parent) block. Its previous hash field always contains all zeroes as it is the first block of the blockchain. Based on the hash value of the genesis block, a chain of blocks is formed. Figure 6.1 shows the structure of the block.

Fields that are part of block are explained as follows:

1. **Block:** Block depicts the block number of the blockchain.
2. **Time Stamp:** Time stamp gives the information of the time when this block is being mined. It contains the Unix time, which is universal and is mostly used for programming purposes. This helps in updating the information of the block after every single second. Time stamp also helps in generating the nonce range.
3. **Nonce:** Nonce is number used only once. Nonce gives extra flexibility of generating hash as everything else is fixed; only nonce can be varied to create the desired hash. Nonce is just a number that can range up to 4 billion. The task of varying nonce is done by miners.
4. **Data:** Data of the block consists of the list of transactions. Transactions are picked from mempool, which stores the unconfirmed transactions.
5. **Previous Hash:** Previous hash helps maintain the cryptographic chain of blockchain as hash of the previous block must match with the previous hash field of the current block. This makes the chain immutable as no one can change the data of the block. Even if someone tries to tamper with the data, its current hash will change and will no longer match with the previous hash field of the next block; hence, it becomes practically impossible to make any changes over the block.
6. **Hash:** Hash is similar to the fingerprint of the block. The SHA-256 hashing algorithm is used to create hash of the block. SHA-256 possesses the satisfying requirements of generating the block, i.e., it is one-way, deterministic,

computationally faster, and withstands collision. With one small change inside the block, its hash value is affected a lot, which is its most important feature.

6.7 BLOCKCHAIN COMPUTATION

Smart devices or mobile devices generate data that require computations for further processing. The computed data is stored in the blocks and computations are also done for verifying the content of the blocks. Figure 6.2 represents the different stages of transferring the data into the block.

1. **Initialization of Data:** Data originates from smart devices or mobile devices. Generally, these devices do not have much capability to do computations. To provide security features to these devices, blockchain plays a very important role. The type of data produced depends on the application that produces that data.
2. **Generation of New Block:** Log events that are part of records can perform operations such as reading/writing on the existing record or creating a new record. These logs are compiled, hashed, and finally, become part of the new block.
3. **Consensus Mechanism Verifies the Block:** The consensus algorithm provides safety and effectiveness to the blockchain. All blocks of the blockchain are of equal status. Such blocks reach consensus by implementing the prior consent of the rules. A few consensus algorithms are PoW, Proof of Stake (PoS), Delegated Proof of Stake (DPoS), Practical Byzantine Fault Tolerance (PBFT), Simplified Byzantine Fault Tolerance (SBFT), Proof of Elapsed Time (PoET), Proof of Relevance (PoR), and Proof of eXercise (PoX).
4. **Append Block into the Ledger:** As the block gets verified by the consensus algorithm and also by other miners that are part of mining pool, the block becomes valid to be appended into the blockchain.

6.8 VARIOUS PLATFORMS FOR THE COMPUTATION OF BLOCK

Smart devices or mobile devices are not very capable of performing computations as they are low-powered. This constraint becomes crucial for the applications of blockchain. Various blockchain technologies use PoW to generate relevant blocks. This process involves solving complex mathematical problem for which huge amounts of computational power is required. It is basically a cryptographic puzzle in which hash must be produced with leading zeroes by varying the nonce field of the block. This cryptographic puzzle is hard to solve as various iterations are done to create nonce,

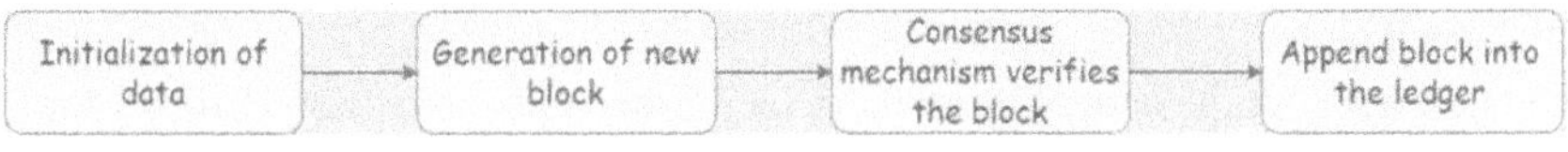

FIGURE 6.2 Creation of the block in the blockchain.

but it is easy to verify as hash can be easily verified using all fields of the block. Due to a lot of computations involved in this consensus algorithm, other consensus algorithms, such as PoS, PoX, PoR, etc., are also used in various platforms. Mining is a process that is used to create secure transactions that are part of blockchain. After the process of mining, blocks are added to the blockchain. This process consumes a lot of computational power. Groups of people who are involved in mining are known as miners. In the case of PoW, miners need to come up with nonce, which produces leading zeroes in hash. The brute force approach is applied for creating a nonce. The miners need to go through various iterations that can produce nonce. A lot of electricity is consumed when producing nonce. Hash produced is the proof of work. Before adding block to a system, the block is verified by all systems in the network. In this way, a lot of computations are required for verifying and validating block. These devices need some platform where blockchain computations can be deployed so that computing power for hashing, applying encryption algorithms, and consensus algorithms like PoW, etc., can be achieved. The data produced by the smart devices is generally offloaded to a blockchain-based network. The most promising approach of the blockchain is the distributed computing. MPC is a category of cryptography that allows a group of mutually distrusting parties to mutually execute computations on the input given to them. Edge nodes that lie very close to the network also have sufficient resources for performing computations that are transferred by the smart devices. A lot of work is done to integrate smart devices with the edge servers for supporting the deployment of blockchain. On the basis of Ethereum, smart contract, and Xtrem Web-HEP, a blockchain-based entirely distributed computing infrastructure known as iEx.ec is projected. Berkeley Open Infrastructure for Network Computing (BOINC) is an open-source middleware platform for network computing that can support blockchain computations.

6.9 ISSUES AND CHALLENGES

The challenges identified during the literature review of blockchain computations are as follows:

1. **Scalability:** The size of blockchain of bitcoin has consistently increased since its creation. Its size was approximately 269.82 GB at the end of March 2020, which increases its bootstrap time as well. Although the internet is massively big, blockchain still suffers from scalability issues because of decentralized networks.
2. **Throughput:** The throughput of blockchain is generally low because it handles fewer transactions compared to other payment processors (Xie, Yu, Huang, Xie, Liu, & Liu, 2019). Blockchains like Bitcoin and Ethereum have the capacity to handle 7 to 20 transactions per second on average, whereas a Visa credit card has the capacity to handle 2,000 transactions per second on average.
3. **Latency:** As the size of blockchain increases, it becomes very difficult for miners to solve the consensus mechanism, and it takes time to generate block. The increase in the time to create block also creates latency issues.

Latency is basically the product of two factors, i.e., computational latency and transmission latency. As blockchain evolves, some acceptance issues may arise because of issues like latency, scalability, etc.

4. **Resource Utilization:** A huge amount of resources are required by miners to solve the cryptographic puzzle and for computation purposes. If resources are utilized optimally, it may reduce the energy requirement of these energy-hungry devices.
5. **Data Integrity:** It is very important to preserve the data integrity of the data. Offloading confidential data to other platforms may violate the integrity of the data. There is a requirement for more reliable data verification techniques for data producers as well as data consumers.
6. **Adaptability:** The number of smart devices and applications using the concept of blockchain is increasing day by day. There should be flexibility for devices to connect or leave the network, and the system must have capability to withstand the fluctuating demands of the users.
7. **Computation:** Solving the consensus algorithms requires high computation power, time, and energy which cannot be efficiently executed on the smart devices. Given the mining process of blockchain is computation-intensive, it is very important to perform the computations in an efficient manner. Smart device computations also need some computing platform as they cannot handle many computations on their end only (Yang, Yu, Si, Yang, Zhang, 2019).
8. **Energy:** Smart devices have restricted computation capacity and battery power. Running the mining process on smart devices may require too much energy consumption, which may restrict the use of blockchain in IoT or mobile environments (Li, Li, Peng, Cui, & Wu, 2019).

6.10 CONCLUSION

Blockchain provides security and privacy measures for various IoT and mobile devices. This chapter elaborates on its various types, features, and applications. For appending a new block into the blockchain, the consensus mechanism plays a vital role. An extensive literature survey has been done to find the integration of blockchain with other technologies for reaching consensus. Various platforms for the computation of newly generated block in the blockchain have been discussed, along with the several issues and challenges. This chapter provides integration of various technologies with blockchain for its efficient computation.

REFERENCES

Altman, E., Reiffers, A., Menasche, D. S., Datar, M., Dhamal, S., Touati, C. (2019). Mining competition in a multi-cryptocurrency ecosystem at the network edge: A congestion game approach. ACM SIGMETRICS Performance Evaluation Review, 46(3), 114–117.

Casado-Vara, R., Chamoso, P., De la Prieta, F., Prieto, J., Corchado, J. M. (2019). Non-linear adaptive closed-loop control system for improved efficiency in IoT-blockchain management. Information Fusion, 49, 227–239.

De Aguiar, E. J., Faiçal, B. S., Krishnamachari, B., Ueyama, J. (2020). A survey of blockchainbased strategies for healthcare. ACM Computing Surveys (CSUR), 53(2), 1–27.

Garcia-Teruel, R. M. (2020). Legal challenges and opportunities of blockchain technology in the real estate sector. Journal of Property, Planning and Environmental Law, 12(2), 129–145. doi: 10.1108/jppel-07-2019-0039.

Guo, S., Hu, X., Guo, S., Qiu, X., Qi, F. (2019). Blockchain meets edge computing: A distributed and trusted authentication system. IEEE Transactions on Industrial Informatics, 16(3), 1972–1983.

Hammi, M. T., Hammi, B., Bellot, P., Serhrouchni, A. (2018). Bubbles of Trust: A decentralized blockchain-based authentication system for IoT. Computers Security, 78, 126–142.

Huh, S., Cho, S., Kim, S. (2017, February). Managing IoT devices using blockchain platform. In 2017 19th international conference on advanced communication technology (ICACT) (pp. 464–467). IEEE.

Jiao, Y., Wang, P., Niyato, D., Xiong, Z. (2018, May). Social welfare maximization auction in edge computing resource allocation for mobile blockchain. In 2018 IEEE international conference on communications (ICC) (pp. 1–6). IEEE.

Kukreja, V., Dhiman, P. (2020, September). A deep neural network based disease detection scheme for citrus fruits. In 2020 international conference on smart electronics and communication (ICOSEC) (pp. 97–101). IEEE.

Kukreja, V., Marwaha, A., Sareen, B., Modgil, A. (2020, June). AFTSMS: Automatic fleet tracking & scheduling management system. In 2020 8th international conference on reliability, infocom technologies and optimization (trends and future directions) (ICRITO) (pp. 114–118). IEEE.

Khan, M. A., Salah, K. (2018). IoT security: Review, blockchain solutions, and open challenges. Future Generation Computer Systems, 82, 395–411.

Kumar, N. M., Mallick, P. K. (2018). Blockchain technology for security issues and challenges in IoT. Procedia Computer Science, 132, 1815–1823.

Lao, L., Li, Z., Hou, S., Xiao, B., Guo, S., Yang, Y. (2020). A survey of IoT applications in blockchain systems: Architecture, consensus, and traffic modeling. ACM Computing Surveys (CSUR), 53(1), 1–32.

Li, J., Li, N., Peng, J., Cui, H., Wu, Z. (2019). Energy consumption of cryptocurrency mining: A study of electricity consumption in mining cryptocurrencies. Energy, 168, 160–168.

Liu, M., Yu, F. R., Teng, Y., Leung, V. C., Song, M. (2018). Computation offloading and content caching in wireless blockchain networks with mobile edge computing. IEEE Transactions on Vehicular Technology, 67(11), 11008–11021.

Lu, Y. (2018). Blockchain and the related issues: A review of current research topics. Journal of Management Analytics, 5(4), 231–255.

Luong, N. C., Xiong, Z., Wang, P., Niyato, D. (2018, May). Optimal auction for edge computing resource management in mobile blockchain networks: A deep learning approach. In 2018 IEEE international conference on communications (ICC) (pp. 1–6). IEEE.

Minoli, D., Occhiogrosso, B. (2018). Blockchain mechanisms for IoT security. Internet of Things, 1, 1–13.

Mingxiao, D., Xiaofeng, M., Zhe, Z., Xiangwei, W., Qijun, C. (2017, October). A review on consensus algorithm of blockchain. In 2017 IEEE international conference on systems, man, and cybernetics (SMC) (pp. 2567–2572). IEEE.

Nguyen, D. C., Pathirana, P. N., Ding, M., Seneviratne, A. (2019). Secure computation offloading in blockchain based IoT networks with deep reinforcement learning. arXiv preprint arXiv:1908.07466.

Nakamoto, S., Bitcoin, A. (2008). A peer-to-peer electronic cash system. Bitcoin. https://bitcoin.org/bitcoin.pdf, 4.

Pan, J., Wang, J., Hester, A., Alqerm, I., Liu, Y., Zhao, Y. (2018). EdgeChain: An edge-IoT framework and prototype based on blockchain and smart contracts. IEEE Internet of Things Journal, 6(3), 4719–4732.

Raikwar, M., Mazumdar, S., Ruj, S., Gupta, S. S., Chattopadhyay, A., Lam, K. Y. (2018, February). A blockchain framework for insurance processes. In 2018 9th IFIP international conference on new technologies, mobility and security (NTMS) (pp. 1–4). IEEE.

Reyna, A., Martín, C., Chen, J., Soler, E., Díaz, M. (2018). On blockchain and its integration with IoT. Challenges and opportunities. Future Generation Computer Systems, 88, 173–190.

Seng, S., Li, X., Luo, C., Ji, H., Zhang, H. (2019, May). A D2D-assisted MEC computation offloading in the blockchain-based framework for UDNs. In ICC 2019-2019 IEEE international conference on communications (ICC) (pp. 1–6). IEEE.

Sharma, P. K., Park, J. H. (2018). Blockchain based hybrid network architecture for the smart city. Future Generation Computer Systems, 86, 650–655.

Xie, J., Yu, F. R., Huang, T., Xie, R., Liu, J., Liu, Y. (2019). A survey on the scalability of blockchain systems. IEEE Network, 33(5), 166–173.

Xu, X., Zhang, X., Gao, H., Xue, Y., Qi, L., Dou, W. (2019). BeCome: Blockchain-enabled computation offloading for IoT in mobile edge computing. IEEE Transactions on Industrial Informatics, 16(6), 4187–4195.

Xiong, Z., Feng, S., Niyato, D., Wang, P., Han, Z. (2018, May). Optimal pricing-based edge computing resource management in mobile blockchain. In 2018 IEEE international conference on communications (ICC) (pp. 1–6). IEEE.

Xue, L., Teng, Y., Zhang, Z., Li, J., Wang, K., Huang, Q. (2017, September). Blockchain technology for electricity market in microgrid. In 2017 2nd international conference on power and renewable energy (ICPRE) (pp. 704–708). IEEE.

Yang, R., Yu, F. R., Si, P., Yang, Z., Zhang, Y. (2019). Integrated blockchain and edge computing systems: A survey, some research issues and challenges. IEEE Communications Surveys Tutorials, 21(2), 1508–1532.

Zhang, Z., Hong, Z., Chen, W., Zheng, Z., Chen, X. (2019). Joint computation offloading and coin loaning for blockchain-empowered mobile-edge computing. IEEE Internet of Things Journal, 6(6), 9934–9950.

Zhu, X., Badr, Y. (2018). Identity management systems for the internet of things: A survey towards blockchain solutions. Sensors, 18(12), 4215.

7 IoT-Inspired Smart Healthcare Service for Diagnosing Remote Patients with Diabetes

Huma Naz, Rishabh Sharma,
Neha Sharma, and Sachin Ahuja
Chitkara University Institute of Engineering & Technology, Chitkara University
Punjab, India

CONTENTS

7.1 INTRODUCTION

In our bodies, the hormone insulin is responsible for transforming starches, sugar, and other aspects of food into needed energy. If our body doesn't form or use insulin, the excessive amount of sugar is evacuated through urination, and this is indicative of a disease referred to as diabetes. Usually, Diabetes Mellitus (DM) occurs when a person has high or above normal blood sugar levels and glucose doesn't reach each body cell. As per the American Diabetes Association [1], 20.8 million children and

DOI: 10.1201/9781003143468-7

adults were diagnosed with this disease in the United States. This impact of diabetes is serious and it can be fatal, and it is related to other medical conditions, such as strokes, blindness, kidney failure, miscarriages, and amputations. Hence, diagnosing diabetes in its early stages plays an essential part in the patient's treatment process, increasing the quality of life. There are three main types of diabetes: type 1, type 2, and gestational diabetes [2]. In type 1 diabetes, the pancreatic cells that are responsible for producing insulin are destroyed; type 1 normally occurs up until the age of 20 years. Type 2 diabetes occurs when the body increases the demand for sugar or becomes insulin resistant. Gestational diabetes most often occurs in pregnant women when a sufficient amount of insulin is not generated by pancreatic cells. To avoid the complications associated with this disease, early detection and treatment is needed. However, data collected from patients directly stored on the cloud server to perform tasks, e.g., diagnosing, accessing health records, predicting the disease, is inadequate for real-time processing. In emergency situations related to the disease, instant medical alerts and quick responses are needed to save lives. Hence, cloud is a centralized repository storage system; and for immediate medical decisions, an intermediate paradigm is required. The IoT cloud-based infrastructure has some drawbacks with latency, delays, availability, distribution, and more. To avoid these kinds of issues, Fog Computing (FC) integrates as a fog layer that acts as a transitional layer between end devices and cloud [3]. Fog makes communication, processing, and transmission faster instead of the cloud-edge based infrastructure. Multiple fog nodes are deployed in the infrastructure closer to the end devices as well as cloud and get responses without any jitter or delay [4]. With predictions, this might be irrelevant if no applicable mechanism can be adapted. Therefore, DM approaches and Machine Learning (ML) algorithms are used to reduce transmissions or delay time and cost for better prediction. The DM technique provides better predictions in areas such as healthcare, academics, and many more [5, 6]. In this chapter, fog-aided DM and ML approaches, along with supporting literature, are discussed. The aim is to diagnose diabetes patients in the early stages to monitor their health using FC delays, bandwidth, and cost of resources.

In any nation's development and progress, healthcare services are considered an important area that must be managed effectually and competently. Nowadays, an emerging need arises for the instant analysis of users' data as well as real-time decisions without delays. So, cloud, IoT, and fog technology can be combined to develop many smart healthcare applications, which can perform better than the existing healthcare systems.

FC is another emerging technology that brings the cloud services closer to the "Thing," such as sensors, mobile phones, and embedded system. Fog technology places the computing and processing power closer to the IoT devices or mobile phones as compared to cloud computing that reduces latency and communication overhead. This is because the physical distance between IoT devices and fog nodes is shorter, and it takes a minimum response time for real-time decision making. FC also performs the computation of highly sensitive data on a local gateway, which improves the security of sensitive data [7]. Due to local computation, FC reduces the load from more centralized resources, which leads to less congestion on the network.

This chapter is organized in various sections. Section 7.2 presents the background of cloud computing, IoT, and FC. A review of related literature is presented in

Section 7.3. Section 7.4 surveys the healthcare mechanism based on diabetes followed by cloud, IoT, and fog-based healthcare frameworks applications. The results are discussed in Section 7.5. This chapter concludes with Section 7.6.

7.2 BACKGROUND OF CLOUD COMPUTING, IoT, AND FC

In this section, an introduction to cloud computing, IoT, and FC is provided.

7.2.1 Cloud Computing

Cloud computing is an emerging technology that provides cost-effective IT resources such as infrastructure, storage, servers, applications, and services to end users. It offers easy-to-use software and data through the internet that can be shared among different clients, and requires no technical expertise. In the last few years, cloud computing has gained popularity due to its significant features and benefits like a lower investment upfront, low operating costs, fast deployment, dynamic scalability and automated self-provisioning of resources, architecture abstraction, pay-as-you-go model, low-cost disaster recovery, massive data storage solutions, easy access and fewer maintenance expenses, ubiquity (i.e., device and location independent), and operational expense model. One of the most significant features of cloud computing is its ability to store enormous amounts of data from diverse business establishments that can be shared among them. Cloud computing provides three importance services: infrastructure as a service (IaaS), platform as a service (PaaS), and software as a service (SaaS), respectively as shown in Figure 7.1. IaaS offers support for hardware, storage, software, servers, and other infrastructure components in the form of virtualized computing resources to end users over the internet on a pay-per-use basis [8]. PaaS is a cloud service provider that provides hardware and software services to its users for the development of application and software. A PaaS provider hosts both the software and hardware applications to its users on its own infrastructure. As a result, PaaS provides a service in which users don't need to install or run hardware and software applications. SaaS is a software distribution model, which hosts the cloud services, making them available to customers over the internet. Among these services, we have especially seen a dramatic growth with database as a service (DBaasS), in which enterprises are allowed to store their databases in the cloud and to access them through internet. The data stored in cloud is considered data outsourcing since it is managed by an external party. The DBaaS model is a major shift from the traditional model of data management that provides seamless mechanisms for organizations to operate their databases upon, from anywhere in the world using internet. Outsourcing databases into the cloud suggests great suitability to enterprises since they don't have to maintain the complexities of direct hardware management and database management on their local machines. Today, many cloud-based databases services, such as Microsoft Azure SQL, Amazon SimpleDB, Google App Engine, and ClearDB, operate entirely in cloud for customers and enterprises.

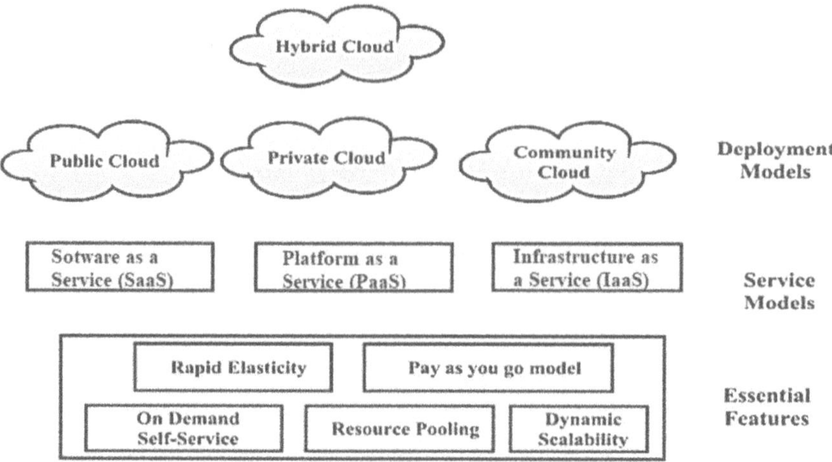

FIGURE 7.1 Cloud computing architecture.

7.2.2 IoT

IoT was originally referred to as the "embedded internet." The acronym IoT was first coined by Kevin Ashton in 1999 [9]. IoT refers to the network of physical objects or things that are embedded and connected with software or sensors by means of exchanging and sharing data with one another. Since IoT came into picture, it is utilized in almost all home automation, smart healthcare applications, smart transport, and utilities applications [10]. There are different key-enabling technologies available for IoT comprising Radio-Frequency Identification (RFID), smart detection, wireless sensor network technology, artificial intelligence, and cloud computing [11]. IoT is an advanced, rapidly developed network of internet-connected sensors that are embedded in a wide-ranging variety of applications [12, 13]. The idea of advancement in the IoT came into existence after a decade; furthermore, development of low-power sensor technologies, connectivity, cloud computing platforms, and ML also enhances its capability.

7.2.3 Fog Computing

A distributed computing paradigm approach was introduced by CISCO to overcome the flaws of cloud computing. FC enables users to access services and computation between cloud and end users closer to each other without any traffic congestion, without delay and security issues, and without depending on the internet. As large amounts of data are generated from the IoT, sensors, and smart wearable devices to monitor the gathered data on cloud, decentralized computing is required. A fog-based layer is integrated with the cloud infrastructure, in which applications, storage, data analytics, and objects are deployed between the fog nodes and cloud storage. Multiple fog nodes are also deployed according to the infrastructure and network requirements. The key idea of introducing FC is to conquer the restrictions and obstructions that arise through the cloud while processing, executing, or replying to requests on a network. Hence, the FC paradigm is beneficial in aspects of latency, bandwidth, delays, jitters, fault tolerance, security, responsiveness, energy, and speed. It can be said that FC boosts up the overall efficiency of the network. A scenario is introduced in Figure 7.2, in which a fog layer is located between the IoT devices and cloud computing to make cities smarter. Likely in smart

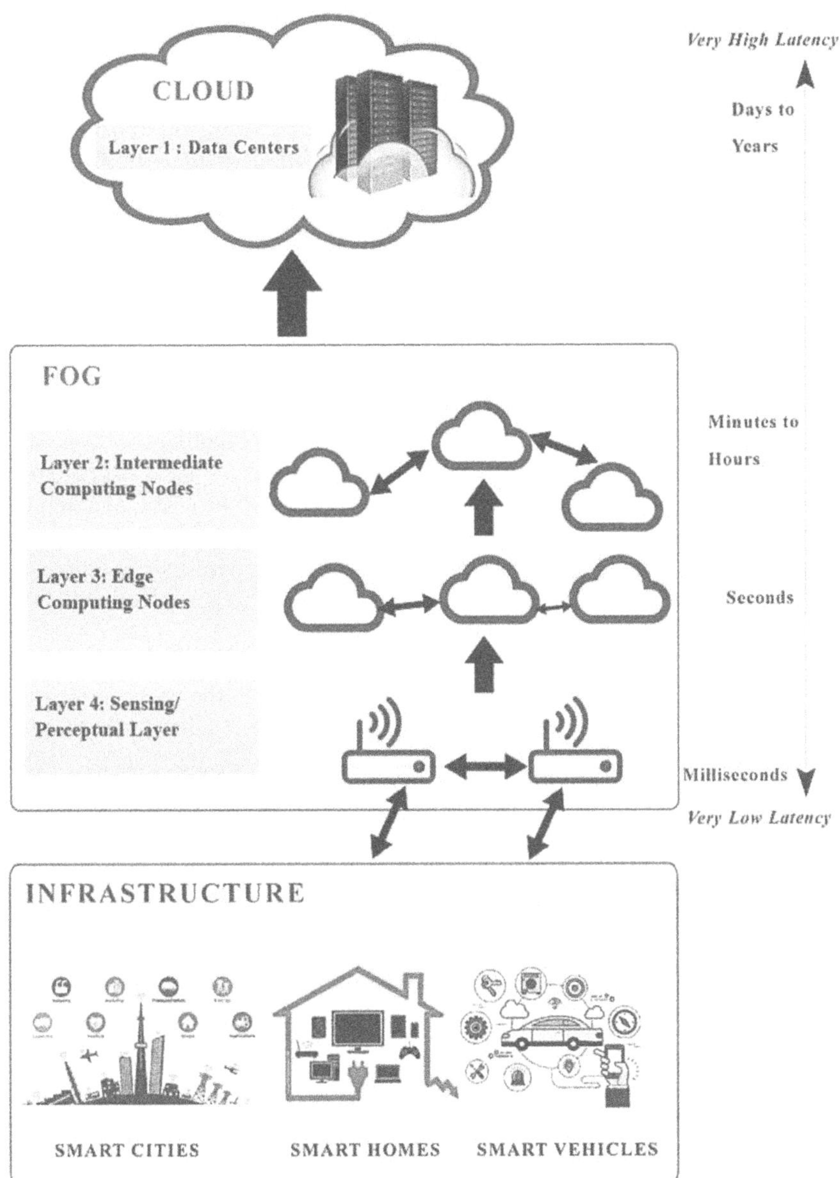

FIGURE 7.2 Fog computing architecture.

cities, a lot of IoT-driven applications, such as a traffic light and management system, smart buildings, health management, parking systems, sanitation, among others, are latency-based and the data generated by these devices is a huge volume that requires huge bandwidth, low latency rates, and minimum response times and costs. But the cloud does not fulfill these mentioned requirements. In order to overcome the cloud drawbacks, FC accomplishes these issues.

7.3 RELATED WORK

Utility DM techniques have outmoded the extant technologies in terms of accuracy and accurate prediction. DM techniques and ML are efficient in extracting hidden and concealed patterns that may result in true information and remain undetected while using other statistical and analytical methods. ML techniques are capable of training the data at a perfect level, so that the outcome will be more likely to be precise while testing. On the other hand, FC decentralizes the data center's resources and makes them closer to the user for improving the user experience and quality of service. This method of computing works as an intermediate node; therefore, computational resources are not required from cloud data centers. The subsequent literature review shows the usability of DM and ML techniques around FC that may have better outcomes for the prognosis of chronic diseases like diabetes.

According to Rabindra Kumar Barik et al. [14], mist computing is proven to be functional for medical health applications. In this end, devices aided the fog nodes to decrease the latency and augment supporting the edge of the client, and improve the client experience. In this paper, the rise of mist computing has been discussed for DM analytics from healthcare applications. This paper recommends and proposes the mist computing-based framework, i.e., mist learn used to detect patients suffering from the DM in the real world.

In essence, Rabindra Kumar Barik et al. [15] proposed a fog-based framework that is used for diabetes prognosis, i.e., fog learns along with the application of this methodology. The proposed methodology implements the Ganga River management approach and K-means clustering technique for detecting the diabetic patients through real-world data. It also implements ML techniques for extracting the pathological feature obtained from the smart watch worn by the diabetic patient. The result shows the predicted outcome of the disease with an analysis of medical and real-world big data of the diabetic patient through FC.

Rojalina Priyadarshini et al. [16] proposed an FC-based neural network deep fog that can analyze the detection and prediction of diabetes, types of stress, and hypertension. This given methodology and the deep fog framework process and group data from individuals and realize the fitness of a particular FC node based on neural networks that can easily handle the mixed and complex data. The proposed system and architecture validated the outcomes and effectiveness for the precise monitoring of critical disease and fitness levels.

In 2017, Gunasekaran Manogaran et al. [17] proposed the IoT framework and use of big data for smart and secure tracking. Smart wearable watches produce a huge amount of data, including structured and unstructured data, which further results in big data, and it is tricky to extract meaningful, constructive, and functional data from big data. To solve this issue, IoT-based architecture has been proposed that consists of two sub-parts, named as metadata for redirecting, and grouping for selecting architectures. These architectures come together to secure the system by integrating FC and cloud computing. Furthermore prediction of the disease is done through the map-based prediction model. A performance evaluation of this proposed framework has been done with consideration to throughput, effectiveness, efficiency, sensitivity, and F-measures.

IOT technology implements the data in an efficient way to reach out to remote and mobile healthcare patient monitoring locations. Despite its reimbursement, this produces extraordinary data that can be easily handled through cloud computing platform. Therefore, Prabal Verma and Sandeep K. Sood [18] proposed the IOT implementation for health monitoring in smart homes and used the FC framework. This framework can overcome the problems by working as an intermediate node. Some advanced features, such as the data warehouse, appropriate data storage, and a reply service at the boundary of the smart home system have also been introduced.

Ahmed et al. [19] introduced a model in 2016 as a fog-based healthcare monitoring model, named Health fog. In their proposed work, the fog layer is introduced as an intermediator in the middle of the cloud and edge devices. For enhancing safety, a cloud access security broker (CSAB) is incorporated into Health fog along with the integration of cryptographic primitives for securing data to be sent through cloud. IoT plays an essential role in the remote health monitoring systems. Therefore, the Ahmed et al. presented an architectural-based IoT-based health as a u-healthcare monitoring system that can provide services to remote patients. This architecture mainly focuses on the IoT-based service delivered to the edge of smart homes and smart hospitals, so that the end user can get all the healthcare services without difficulty. In 2018, Negash et al. [20] took this tour on continuation and focused on an FC-based smart e-health gateway implementation for reaching out to remote patients. They thought of the implantation of gateways as the connection between home and hospital that can connect network. Moreover, the network functions and parameters are discussed.

IoT-based healthcare applications require immediate analysis of health-oriented data generated by IoT devices for making real-time decisions and generating recommendations for a healthy lifestyle. Since IoT is capable of providing advance health diagnoses and medication, Latif et al. [21] proposed a novel approach with wireless sensors networks and cloud-based servers for patient's continuous health status monitoring. The methodology consists of a wearable health monitoring system, an AI-based tool, cloud big data storage, and an analytic prognostic system and medicine-dispensing system. Few tests have been done on system to check if the system achieved the intended objective, including providing a continuous health-monitoring status and timely dosages to the patient. The proposed system helps healthcare officials to view the effects of medication on patients and remotely monitor their health status.

Medicinal services are fundamental to humans. It is crucial for the advancement and improvement of a society. In recent years, IoT has played a very major in the revolution of healthcare industry. Bhatia et al. [22] proposed an effective and centric urine-based diabetes monitoring system. Mainly, the system contains four stages for a systematic diagnosis of a urine-based diabetes infection. The four stages include the data acquisition stage, data classification stage, feature extraction/mining stage, and diabetic prediction stage. The prediction of diabetes is done using recurrent neural network (RNN). The experiments are validated on four different individuals. With this knowledge, Akkas et al. [23] discussed advancements in healthcare and patient monitoring using IoT. They further presented applications of IoT technology in diverse medical fields and some future trends with Bio-IoT or Nano-IoT systems. According to the authors, the most important component of a remote patient's health monitoring is Wireless Body Area Network (WBAN), which is placed on the body of a patient and can communicate wirelessly. The methodology presents the implementations of a biomedical application based on WBAN. The collected data is transmitted from a diverse wireless sensor network using IoT devices.

Data mining with IoT emerged as an influential computing technique in the healthcare industry. Researchers have used different data mining techniques with IoT for diabetes and other disease detection so far. However, more efficient techniques are needed for the detection of diabetes and heart disease. Therefore, a smart data mining-IoT enabled advanced technique has been proposed for the early detection of diseases [24]. This technique comprises bio sensors, IoT, chatbots, semantic analysis, and granular computing. Bio sensors help get the data of concerned patients. As discussed, IoT has a lot of advantages in the field of healthcare; thus, Fradin et al. [25] proposed an IoT-based and cloud-based healthcare diagnosis system with the ML algorithms. In the proposed system, the data are recorded through wearable sensors; then the processed signals/data are transmitted to the network in cloud environment. The article also presents the novel hybrid approach of decision tree. In this process, a new feature set is created for testing neural fuzzy model. The diagnosis of specific disease or diabetic problems can be simulated with efficient outcomes. A survey on security of IoT framework was performed by Ammar et al. [26]. In this survey, eight main frameworks of IoT are considered and a detailed comparative analysis is performed considering their proposed architecture, issues in the development of third-party smart applications, and hardware and software compatibility for ensuring security. With the growing populations and older people, the society needs a personalized healthcare system to avoid and manage the chronic condition. Therefore, Wang et al. [27] discussed three key points in the paper. The first is to review the key factors of the home-based remote healthcare system; the second is to present the latest advances of remote healthcare system; and the third is to review the recommendations for home-based in-home healthcare monitoring.

FC is another emerging technology that brings the cloud services closer to the "Thing," i.e., sensors, mobile phones, and an embedded system. Thus, Devarajan et al. [28] proposed a fog-assisted approach to maintain the glucose level. The decision tree classifier is used to predict the risk level of diabetes for achieving high classification accuracy. With FC, an emergency alert is generated for preventive measures. Experimental results illustrated an improved accuracy, computational complexity, and latency.

The term IoT for the most part alludes to situations where organized networks and registrants' ability reach out to objects, sensors, and empower these gadgets to create, trade, and devour information with little human intervention, utilizing different systems' administration and correspondence models. Over the years, IoT is used in diverse areas such as medical and healthcare systems, smart homes, remote services, and many other areas [29]. The authors in [29] present the different applications of IoT; moreover, the possible potential of the technology in the future trends is also presented.

7.4 METHODOLOGY

To accomplish the specified aim, a framework consisting of a three-tier architecture – namely, the end layer, fog nodes layer, and cloud computing – is proposed in this chapter. With the help of deep learning, accurate results are attained on the end layer. After that, the data collected by end devices is sent to the fog server and then to the cloud. The dataset applied in this dataset is a Pima dataset and a fog layer integrated with the cloud infrastructure helps to achieve results without any delay or jitters.

7.4.1 HEALTHCARE APPLICATION

In healthcare applications, IoT-based sensors like electrocardiography (ECG) sensor, heart rate monitor, Global Positioning System (GPS) sensor, Radio-Frequency Identification (RFID) sensor, accelerometer, and blood pressure monitor sensor are the major sources for acquiring data such as the vital signs of the user and surrounding environment parameters. The data generated from the sensors can be captured using mobile devices. The information collected from sensors is combined with a user's personal information, such as name, age, gender, weight, and occupation. Mobile phones play an important role in the healthcare applications and can be integrated with IoT devices and cloud servers for expanding the network. The data generated by applications is stored on the cloud; thus, realizing the importance and need of remote analysis of patients is extremely significant. Also, the direct analysis of a patient's real-time data and decisions without any delay is in demand nowadays. Health monitoring applications are considered among the fastest growing applications among mobile applications with the integration of cloud computing and IoT. This integration has been proven as a very helpful resource for gathering users' data and plays an essential role in diagnosing and generating recommendations to remote patients for a healthy lifestyle. To further improve the quality of service (QoS) and immediate notifications of real-time applications, there is another trend to provide computing and processing services at the edge of the network devices, i.e., mobile phone, routers, hubs, and switches. To achieve this objective, cloud computing raises major issues, such as transmission latency, a high power consumption rate, location awareness issues, and a degradation of services due to huge volume of traffic between IoT devices and cloud. So, FC can be the glue factor for most of the applications running on real-time data generated by mobile and sensor devices to provide computing and processing services at the edge of network. Chronic diseases such as hypertension, diabetes, and coronary heart disease (CHD) are the leading causes of death. Early detection and prevention of CHD is essential, as doctors recommend that these diseases can be controlled only with the prior knowledge and healthy diet. It can be said that precautionary measures and timely medical treatment can reduce the mortality rate of diabetic patients. It is also the most important challenge for officials, medical officers, and private healthcare agencies to confirm the safety of their citizens from diseases like diabetes and CHD. Presently, doctors are overburdened with patients as diverse, new viruses are on the rise and there is an inadequate number of caregivers in hospitals. Therefore, physically monitoring every patient's case is very tough for doctors. Such type of diseases cannot be effectively predicted by using the existing healthcare systems. To overcome these issues, a fog-assisted cloud-based healthcare system with localization technology can be utilized to track the current location of high-risk patients with chronic diseases and their health behaviors to provide quick service as well as improve the quality of care in hospitals and the home. So, incorporating fog into IoT is required to provide the identification, prevention, and control measures for chronic diseases like diabetes, from remote sites, in the early stages.

7.4.2 Dataset Description

Various experiments have been done using the Pima Indian Dataset (PID). The dataset was originally from the National Institute of Diabetes and Digestive and Kidney Diseases (NIDDK). PID was taken from the UC Irvine Machine Learning (UCI ML) repository for this work [30]. The reason for selecting this dataset is that most of the people in modern times are living with an identical type of lifestyle that includes a high reliance on processed food coupled with declining physical activities. Pima is a group of Native Americans who lived in an area now known as central Arizona. Due to their genetic predisposition, they can survive on low carbohydrates for many years. However, during recent past, the Pima group suddenly shifted from their traditional diet toward processed food, followed by a decrease in their physical activities [31]. Consequently, they were detected with high levels of type 2 diabetes; and for this reason, since 1965, their health data have been used in many studies related to diabetes.

PID includes a certain number of medical predictors and one variable target. The predictor variables are the number of pregnancies, body mass index (BMI), blood pressure, skin thickness, insulin level, age, glucose, and diabetes pedigree function shown in Table 7.1. All the participants in the PID study are females up to the age of 21. The dataset has 768 instances divided into 268 non-diabetic instances and 500 diabetic instances. The target variable identifies whether a person is non-diabetic (represented by 0) or diabetic (represented by 1). The description of different parameters of each attribute in the dataset, including max value, min value, standard deviation, missing value, mean, and the median are given in Table 7.2.

7.4.3 Deep Feed-Forward Neural Network

Deep feed-forward neural networks are commonly called Multilayer Perceptron (MLP) or feed-forward neural networks. A neural network originates from a very

TABLE 7.1
Description of PID Attributes

Sr. No.	Predicators	Description of Predicators	Unit
1.	Pregnancy	Number of times a female participant is pregnant	—
2.	Plasma glucose	Glucose concentration in 2 hours in an oral glucose tolerance test	mg/dl
3.	Diastolic blood pressure	Diastolic blood pressure (upper blood pressure)	mmHg
4.	Triceps skinfold thickness	Skin thickness of participant in mm Concluded by the collagen content	mm
5.	Insulin	Participant's 2-hour serum insulin	mmU/Ml
6.	Body mass index	Weight of a participant in kg/HEIGHT (IN M) ^2)	kg/m^2
7.	Diabetes pedigree function	Appealing attributes used for diabetes diagnosis	—
8.	Age	Age of participants	—
9.	Outcome	Diabetes onset with diabetic and non-diabetic patients	—

famous ML algorithm known as perceptron [32]. Perceptron is a linear classifier with a mapping function that partitions feature space using a linear function that is a boundary line used to classify data into two classes [33]. The MLP neural network is a model with multiple layers of input that are associated with some weight, which are presented in the processor in a feed-forward manner [34], as shown in Figure 7.3.

Figure 7.3 demonstrates the architecture of the feed-forward network where the input layer is denoted as X1 and 1 is the bias value, whereas hidden layers are associated with the activation function represented as F1, F2, and F3. Each input neuron,

TABLE 7.2
Detailed Description of PID Attributes

Sr. No.	Predicators	Missing Values	Mean	Std Dev	Range	Data Type
1.	Pregnancy	0	3.845	3.370	0–17	Integer
2.	Glucose	0	120.89	31.973	0–199	Integer
3.	Diastolic blood pressure	0	316.56	1096.927	0–122	Integer
4.	Skinfold thickness	0	51.697	88.690	0–99	Integer
5.	Insulin	0	819.49	3873.732	0–846	Integer
6.	Body mass index	0	60.769	92.015	0–67.1	Real
7.	Diabetes pedigree function	0	0.472	0.472	0.078–2.42	Real
8.	Age	0	33.241	11.760	21–81	Integer
9.	Outcome	(Diabetic instances – 268) (Non-diabetic instances – 500) (Total instances – 768)				Polynomial

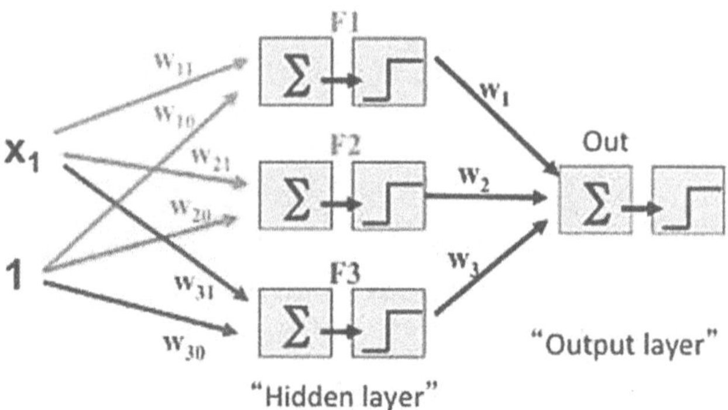

FIGURE 7.3 Multilayer perceptron model.

whether travelling toward the hidden layer or output layer, has some weight connected to it. The processor uses an activation function to produce the output, which is represented as F1, F2, and F3 in the hidden layer's nodes [35]. If the predicted output in the output layer is the same as the desired output, then the performance is considered satisfactory. No change will be made in the weights; otherwise, the weights will be updated to reduce errors.

7.4.4 Pros of Using FC

Cloud computing is in trend from last few years due to its applications in central storage, data processing and analysis, but due to huge amounts of data being forwarded to cloud, traffic congestion increases significantly. Despite this, the computation process takes more energy, which makes the cloud infrastructure a greater energy consumer. On the other hand, with the use of FC data, transmission and computation processes take place locally as fog makes end users closer to the edge of the network [36]. Hence, in emergency cases, FC provides lower energy consumption, bandwidth and low-latency services, and fends off some grievous consequences.

7.5 EXPERIMENTAL EVALUATION

In this purposed framework, DM, along with ML algorithms, e.g., decision trees and neural networks, is adapted to monitor diabetes patients for accurate predictions. The applied data will be sent to fog nodes through the fog layer, which acts as an intermediary between cloud and end users. To evaluate the efficiency of Fog Device Combined Approach (FDC), four interpretation metrics are considered: recall, accuracy, precision, and f-measure. By using these algorithms and mechanisms, we have simulated our results. The outcome of this framework completely proves that introducing the fog layer improves the efficiency and computation of the network infrastructure.

7.5.1 Evaluation Metrics

Different evaluation metrics are used to compute the essence of the proposed prediction model. Evaluation metrics tend to play a significant role in measuring the performance of the prediction model. There can be a case where a model can perform well with a specific evaluation metric, but it can be considered less accurate when any other evaluation metric comes into use. So, the selection of suitable performance metrics is an essential part of the development. Several evaluation metrics are available for evaluating the performance of classification models, such as classification accuracy, sensitivity-specificity, F1-score, receiver operating characteristic (ROC) curve, and area under the curve (AUC), which can be used to measure performance. But the evaluation parameters considered to evaluate the performance of the applied technique are accuracy, sensitivity, and specificity, which are discussed here in detail.

- **Accuracy:** Accuracy of a measurement is the ratio of correctly classified observations to the total number of observations [37], or accuracy can be referred to as the closeness of a measured value to a standard or known value. Accuracy can be measured using Eqn. (7.1). Accuracy is a good performance measure when the target variable class data is nearly balanced.

$$Accuracy = \frac{TP + TN}{TP + FP + FN + TN} \tag{7.1}$$

  Here, true positives are abbreviated as TP; true negatives are abbreviated as TN; false positive are abbreviated as FP; and false negatives are abbreviated as FN.
- **Sensitivity:** Sensitivity is a proportion of the true positive correctly classified [38]. Sensitivity is a performance measure that evaluates the true positives from each class label and can be measured using Eqn. (7.2).

$$Sentivity = \frac{TP}{TP + FP} \tag{7.2}$$

- **Specificity:** Specificity is a proportion of true negative correctly classified [39]. Specificity is an evaluation metrics that measures the true negatives from each class label and can be measured using Eqn. (7.3).

$$Specificity = \frac{TN}{TN + TP} \tag{7.3}$$

TABLE 7.3
Combined Performance Measures of DL Algorithm

	Methods	
Performance Measures	**Proposed Approach (Weka)**	**Proposed Approach (Rapid Miner)**
Accuracy (In %)	96.62	96.23
Precision (In %)	95.06	98.24
Recall (In %)	96.35	97.71
F-Measure (In %)	93.21	99.14
Specificity (In %)	97.86	95.52
Sensitivity (In %)	94.03	96.23

precision, recall, and F-measure, were calculated for the classification algorithm using the PIMA dataset. Table 7.3 shows that our proposed integrated approach outperformed in every performance measure and provided the best result for the detection of diabetes in remote patients with an accuracy rate of 96.13%. Figure 7.4 shows the comparison between the distinct performance metrics for diabetes detection.

As presented in Figure 7.4 and Table 7.2, our proposed approach provided reliable accuracy on the PIMA dataset. The achieved accuracy rate of 96.67% shows that the approach can be used as a prognostic tool for early diabetes detection to remote patients. Those features that do not contribute to the study need to be pruned [40]. In this study, we have used the best-selected attributes for achieving an accurate diagnosis of onset diabetes.

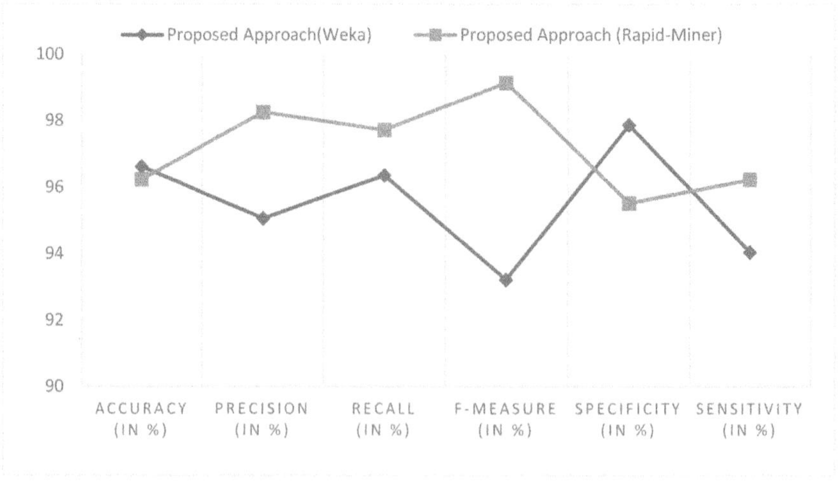

FIGURE 7.4 Comparison of performances of classification method.

7.6 CONCLUSION

A three-tiered cloud-fog based framework consisting of an end layer, fog node layer, and cloud computing is proposed in this chapter. The DL algorithm is applied on the end layer to remotely access the data of healthcare patients. Where users cannot communicate and compute the processes directly with the cloud, a fog layer is introduced in the infrastructure on which user can put their request and perform the computation tasks easily and quickly. An accuracy of 96.67% was achieved on Weka and 96.62% on the rapid miner toolkit. In the future, we will execute this framework with real-time datasets and more parameters, e.g., cost, fault tolerance, security, and delays.

REFERENCES

1. Pham, Huy Nguyen Anh, and Evangelos Triantaphyllou. "Prediction of diabetes by employing a new data mining approach which balances fitting and generalization." *Computer and Information Science*. Springer, Berlin, Heidelberg, 2008. 11–26.
2. Barik, Rabindra K., et al. "FogLearn: Leveraging fog-based machine learning for smart system big data analytics." *International Journal of Fog Computing (IJFOG COMPUTING)* 1.1 (2018): 15–34.
3. Luan, Tom H., et al. "Fog computing: Focusing on mobile users at the edge." *arXiv preprint arXiv:1502.01815* (2015).
4. Bonomi, Flavio, et al. "Fog computing and its role in the internet of things." *Proceedings of the first edition of the MCC workshop on Mobile cloud computing*. ACM, 2012.
5. Kamal, Preet, and Sachin Ahuja. "Academic performance prediction using data mining techniques: Identification of influential factors effecting the academic performance in undergrad professional course." *Harmony Search and Nature Inspired Optimization Algorithms*. Springer, Singapore, 2019. 835–843.
6. Kamal, Preet, and Sachin Ahuja. "An ensemble-based model for prediction of academic performance of students in undergrad professional course." *Journal of Engineering, Design and Technology* 17, no. 4 (2019): 769–781.
7. Stojmenovic, I. "Fog computing: A cloud to the ground support for smart things and machine-to-machine networks." In *2014 Australasian Telecommunication Networks and Applications Conference (ATNAC)* (pp. 117–122). 2014, November. IEEE.
8. Ashton, K. "That 'internet of things' thing." *RFID Journal* 22, no. 7 (2009): 97–114.
9. Atzori, L., A. Iera, and G. Morabito, "The internet of things: A survey." *Computer Networks* 54, no. 15 (2010): 2787–2805.
10. Sundmaeker, H., P. Guillemin, P. Friess, and S. Woelfflé. "Vision and challenges for realising the Internet of Things." *Cluster of European Research Projects on the Internet of Things, European Commission* 3, no. 3 (2010): 34–36.
11. Buckley, J. 2006. "The internet of things: From RFID to the next-generation pervasive networked systems."
12. Atzori, L., A. Iera, and G. Morabito. "The internet of things: A survey." *Computer Networks* 54, no. 15 (2010): 2787–2805.
13. Gubbi, J., R. Buyya, S. Marusic, and M. Palaniswami. "Internet of Things (IoT): A vision, architectural elements, and future directions." *Future Generation Computer Systems* 29, no. 7 (2013): 16451660.
14. Barik, Rabindra Kumar, et al. "Leveraging machine learning in mist computing telemonitoring system for diabetes prediction." *Advances in Data and Information Sciences*. Springer, Singapore, 2018. 95104.

15. Barik, Rabindra K., et al. "FogLearn: leveraging fog-based machine learning for smart system big data analytics." *International Journal of Fog Computing (IJFOG COMPUTING)* 1.1 (2018): 15–34.
16. Priyadarshini, Rojalina, Rabindra Barik, and Harishchandra Dubey. "DeepFog: Fog computing-based deep neural architecture for prediction of stress types, diabetes and hypertension attacks." *Computation* 6.4 (2018): 62.
17. Mohammed, M. N., S. F. Desyansah, S. Al-Zubaidi, and E. Yusuf. "An internet of things-based smart homes and healthcare monitoring and management system." *Journal of Physics: Conference Series* 1450, no. 1 (2020): 012079. IOP Publishing.
18. Verma, Prabal, and Sandeep K. Sood. "Fog assisted-IoT enabled patient health monitoring in smart homes." *IEEE Internet of Things Journal* 5.3 (2018): 1789–1796.
19. Ahmad, Mahmood, Muhammad Bilal Amin, Shujaat Hussain, Byeong Ho Kang, Taechoong Cheong, and Sungyoung Lee. "Health fog: A novel framework for health and wellness applications." *The Journal of Supercomputing* 72, no. 10 (2016): 3677–3695.
20. Negash, Behailu, Tuan Nguyen Gia, Arman Anzanpour, Iman Azimi, Mingzhe Jiang, Tomi Westerlund, Amir M. Rahmani, Pasi Liljeberg, and Hannu Tenhunen. "Leveraging fog computing for healthcare IoT." *Fog Computing in the Internet of Things*, Springer, Cham, 2018. 145–169.
21. Latif, Ghazanfar, Achyut Shankar, Jaafar M. Alghazo, V. Kalyanasundaram, C. S. Boopathi, and M. Arfan Jaffar. "I-CARES: Advancing health diagnosis and medication through IoT." *Wireless Networks* 26, no. 4 (2020): 2375–2389.
22. Bhatia, Munish, Simranpreet Kaur, Sandeep K. Sood, and Veerawali Behal. "Internet of things-inspired healthcare system for urine-based diabetes prediction." *Artificial Intelligence in Medicine* 107 (2020): 101913.
23. Akkaş, M. Alper, Radosveta Sokullu, and H. Ertürk Çetin. "Healthcare and patient monitoring using IoT." *Internet of Things* 11 (2020): 100173.
24. Sharma, Manik, Gurvinder Singh, and Rajinder Singh. "An advanced conceptual diagnostic healthcare framework for diabetes and cardiovascular disorders." *arXiv preprint arXiv:1901.10530* (2019).
25. Abdali-Mohammadi, Fardin, Maytham N. Meqdad, and Seifedine Kadry. "Development of an IoT based and cloud-based disease prediction and diagnosis system for healthcare using machine learning algorithms." *International Journal of Artificial Intelligence* 2252, no. 8938 (2020): 8938.
26. Ammar, Mahmoud, Giovanni Russello, and Bruno Crispo. "Internet of Things: A survey on the security of IoT frameworks." *Journal of Information Security and Applications* 38 (2018): 8–27.
27. Philip, Nada Y., Joel JPC Rodrigues, Honggang Wang, Simon James Fong, and Jia Chen. "Internet of Things for in-home health monitoring systems: Current advances, challenges and future directions." *IEEE Journal on Selected Areas in Communications* 39, no. 2 (2021): 300–310.
28. Devarajan, Malathi, V. Subramaniyaswamy, V. Vijayakumar, and Logesh Ravi. "Fog-assisted personalized healthcare-support system for remote patients with diabetes." *Journal of Ambient Intelligence and Humanized Computing* 10, no. 10 (2019): 3747–3760.
29. Manogaran, Gunasekaran, et al. "A new architecture of Internet of Things and big data ecosystem for secured smart healthcare monitoring and alerting system." *Future Generation Computer Systems* 82 (2018): 375–387.
30. "PIMA Indian Dataset Source" [Online]. Available: http://archive.ics.uci.edu/ml/datasets/Pima+Indians+Diabetes.
31. "Reason of Choosing PIMA Indian Dataset," [Online]. Available: https://www.andreagrandi.it/2018/04/14/machine-learning-pima-indians-diabetes/.
32. Miikkulainen, R., J. Liang, E. Meyerson, A. Rawal, D. Fink, O. Francon, and B. Hodjat. (2019). "Evolving deep neural networks." *Artificial Intelligence in the*

Age of Neural Networks and Brain Computing, 293–312. https://doi.org/10.1016/b978-0-12-815480-9.00015-3
33. Pandi, A., M. Koch, P. L. Voyvodic, P. Soudier, J. Bonnet, M. Kushwaha, and J.-L. Faulon. "Metabolic perceptrons for neural computing in biological systems." *Nature Communications* 10, no. 1 (2019): 1–13. https://doi.org/10.1038/s41467-019-11889-0.
34. Tajmiri, S., E. Azimi, M. R. Hosseini, and Y. Azimi. "Evolving multilayer perceptron, and factorial design for modelling and optimization of dye decomposition by biosynthetized nano CdSdiatomite composite." *Environmental Research* 182 (2020): 108997. https://doi.org/10.1016/j.envres.2019.108997.
35. Kukreja, V. and P. Dhiman. "A deep neural network based disease detection scheme for citrus fruits." In *2020 International Conference on Smart Electronics and Communication (ICOSEC)* (pp. 97–101). 2020, September. IEEE.
36. Goyal, M., R. Goyal, P. Venkatappa Reddy, and B. Lall. "Activation functions." *Deep Learning: Algorithms and Applications* (2019): 1–30. https://doi.org/10.1007/978-3-030-31760-7_1.
37. Hung, A. J., J. Chen, and I. S. Gill. "Automated performance metrics and machine learning algorithms to measure surgeon performance and anticipate clinical outcomes in robotic surgery." *JAMA Surgery* 153, no. 8 (2018): 770. https://doi.org/10.1001/jamasurg.2018.1512.
38. Kukreja, V., D. Kumar, and A. Kaur. "GAN-based synthetic data augmentation for increased CNN performance in vehicle number plate recognition." In *2020 4th International Conference on Electronics, Communication and Aerospace Technology (ICECA)* (pp. 1190–1195). 2020, November. IEEE.
39. Tschandl, P., N. Codella, B. N. Akay, G. Argenziano, R. P. Braun, H. Cabo, and H. Kittler. "Comparison of the accuracy of human readers versus machine-learning algorithms for pigmented skin lesion classification: An open, web-based, international, diagnostic study." *The Lancet Oncology* 20, no. 7, (2019): 938–947. https://doi.org/10.1016/s1470-2045(19)30333-x.
40. Nandyala, Chandra Sukanya, and Haeng-Kon Kim. "From cloud to fog and IoT-based real-time Uhealthcare monitoring for smart homes and hospitals." *International Journal of Smart Home* 10.2 (2016):187–196.

8 Segmentation of Deep Learning Models

Prabhjot Kaur and Anand Muni Mishra
Chitkara University Institute of Engineering & Technology, Chitkara University
Punjab, India

CONTENTS

8.1 INTRODUCTION

An image is a means of transmitting data, and there is plenty of valuable information in the image. Getting important data from an image or recognition of data is done through digital image technology. With the help of image segmentation, a picture can be understood easily [1]. Image segmentation is a basic work of manipulating pictures, interpreting images, understanding images, and identifying patterns. Extracting the field that users are interested in is done using calculations based on parameters in which the same type of data is grouped. This includes clustering images into different fragments or entities. In a vast variety of technologies, including medical image processing, autonomous vehicles, video monitoring, and virtual reality, segmentation plays a central role [2, 3]. There is no common standard procedure for image segmentation because different types of images require

DOI: 10.1201/9781003143468-8

different partitions to extract significant features. On the other hand, when segmenting a particular form of an image, various approaches are not equally efficient, and the parameters for determining a satisfactory segmentation depend on the desired objective of the segmentation itself. Therefore, there is no specific outcome to the segmentation dilemma [4].

Segmentation approaches have been improved and expanded by advancements in computer science and mathematical simulations, and have retained close relationships with other computer vision methods, such as image recognition and edge detection, while differentiating their purposes from them. These approaches solve various types of problems, considering their close relationships, and generate different outcomes. In Sections 8.1.1–8.1.4, we will define four main computer vision issues, which are listed in increasing order of complexity.

8.1.1 Image Classification

The first issue is to classify the group of key objects inside an image.

8.1.2 Object Detection

For any known object within an image, classify the object category and locate its location using a feature vector.

8.1.3 Semantic Segmentation

For all objects within an image, define the object type of every pixel.

8.1.4 Instance Segmentation

In the image segmentation task, the detection of a particular object before classification is known as instance segmentation.

This technique can be conceived as the difficulty of pixel labeling with semantic identification (semantic segmentation) or division of single objects (instance segmentation). Semantic segmentation performs pixel-level labeling for all the pixels of an image with a set of objects that are in an image; it is usually a more difficult undertaking than image classification, which foresees a single identification for the whole image. Other than this, in the instance segmentation, the region of interest are detected and classified according to the objects in the image (e.g., the division of every single object) [5].

To diagnose an infected part from any image, segmentation is the best step at the time of pre-processing. Different techniques of segmentation are used that are either based on intensity (histogram intensity-based segmentation) or pixels (index measure, over lab coefficient). These methods give an accuracy of 98.025% for the segment of the affected region [6]. Based on the clustering of an image dataset fusion, different clustering algorithms are used for the segmentation of diseased part from an image. Then three color features are extracted using a pyramid of histograms of orientation gradients (PHOG) algorithm. Accuracy achieved for the apple disease segmentation is 90.43% and for the cucumber plant is 92.15% [7]. The identification and recognition of disease in citrus fruits is done using a hybrid approach. The principal component analysis (PCA) score, entropy, and skewness-based covariance hybrid approaches are used for extracting the features of citrus disease, such as black spots, cankers, scabs, greening, etc. The classification accuracy achieved is 97% [8]. The newly improved multichannel selection-based segmentation model Chan-Vese (C-V) is used for segmenting the wheat lesion from an image. From the multichannel (R, G, B, H, S, V), only the Red, Green, and Blue (RGB) channel is selected. Different segmentation methods are used with the C-V model in which the k-means segmentation method gives a better result. The accuracy achieved by k-means method is 84.11%, which also calculates performance in terms of efficiency and robustness [9].

Finding and dividing diseased parts from healthy parts using a program is the main challenge when analyzing a plant. Objects and boundaries (lines, curves, etc.) in images are usually located using image segmentation. Some of the parameters used for measuring the performance of the segmentation are Jaccard, dice, variation index, and consistency error method. When compared with different segmentation algorithms, the discrete wavelet transform method with k-means clustering gives a better result [10]. For comparing signs and symptoms in a plant leaf, different algorithms are proposed in computer vision. Some of the steps in this segmentation algorithm are automatic; the difficult one is the selection of the H-component. The novel algorithm-based semi-automatic method allows for versatility without sacrificing speed. Variations in leaf color, symptom color, light disparities, and other factors are all taken into account by the algorithm. The majority of errors were caused by color channel constraints, and some sources of error were unavoidable even with a completely manual approach [11]. Some of the papers consider only a single part of the eye image. Research on multiple parts of the eye image is not focused. The authors of reference number 12 propose a model for calculating the multiple regions of an eye image using different segmentation algorithms [12]. Early detection of any disease will help with curing a person. In this case, the skin cancer disease melanoma can have dangerous effects on human skin. For the detection and segmentation of disease, different segmentation techniques are used to help treat a person at an early stage [13].

Brain diagnosis relies heavily on the segmentation of tissues from medical images. Since manual segmentation is time-consuming, it is essential to construct a program for the segmentation of data using Deep Learning (DL) networks. It's still a challenging task to segment for medical images. Techniques such as cropping and normalisation are used to process the MRI images. Then, for segmentation, an Fully Convolutional Network (FCN)-based deep learning model is built. Finally, the performance of images is evaluated and they are uploaded to the Penn Imaging Website so that model performance can be evaluated [14]. The segmentation of magnetic resonance images is carried out using a differential evolution of the linear population size reduction (LSHADE) method. The proposed method is carried out in different steps: first, the image is converted into grayscale, then it is grouped into magnetic resonance, and finally it forms a group of unhealthy images. The output of the proposed method is compared with machine learning using a metaheuristic approach [15]. Table 8.1 represents a few literature surveys for the segmentation of deep learning models.

TABLE 8.1
Literature Surveys

Ref No.	Year of Publication	Author's/Authors' Name(s)	Type of Dataset	Summary
[16]	2016	Jayme Garcia Arnal Barbedo	Plant leaf	Compared with other different methods, the method proposed in this paper gives a more accurate result. To differentiate the leaf's healthy part from the unhealthy part, this method gives effective result.
[17]	2018	Shu H., Fan K. et al.	Cotton leaf	Proposed an automatic segmentation technique with gradient and local information. Compared with different segmentation techniques, such as the Generalized Arc Consistency (GAC) algorithm, C-V algorithm, and Local Best Fit (LBF) algorithm, the automatic segmentation gives a better result. This model not only segments images of cotton leaves in temperate areas but also can provide technical assistance in accurate cotton disease diagnosis and identification.
[18]	2019	Vijai Singh	Sunflower leaf	Proposed a model for segmentation using Particle Swarm Optimization (PSO), which does not require any extra information for segmentation of the disease image. For better accuracy, the hybrid PSO model is used for the segmentation process.
[19]	2019	Praveen Kumar J. Dominic S.	Plant leaf	Based on segmentation, different steps are followed for counting and segmenting the data. Steps include the statistical-based technique, the graph-based method, and finally, the Circular Hough Transform, which achieves an accuracy of 95.4%.
[20]	2019	Li J., Zhang L. et al.	Tomato leaf	For the clustering of tomato leaf image, different adaptive algorithms are used for the segmentation of images. The clustering values are calculated by the Davies-Bouldin index with clustering calculations. The F1 measure and entropy value are calculated for measuring the segmentation accuracy.
[21]	2020	Riehle D., Reiser D., Griepentrog H.	Maize, Sugar beet plant	Proposed a robust model that segments the image under different conditions, such as overexposure and underexposure. The model relies on an index-based semantic method with an accuracy of 97.4%. This proposed index-based model is more accurate than the other index-based methods.

This chapter explains various segmentation algorithms for analyzing images of any size or frame. The algorithms are categorized according to usability and analysis. The threshold segmentation and regional growth segmentation come under region-based segmentation. Next is the edge-detection segmentation; under this, the Sobel operator and the Laplacian operator work. Another category is the segmentation based on clustering; under this, k-means clustering works.

8.2 OVERVIEW

8.2.1 Region-Based Segmentation

8.2.1.1 Threshold Segmentation

The threshold segmentation method is one of the easiest and most common methods of linear segmentation used for image segmentation. A basic segmentation algorithm is used under this approach, through which processing the information in the images can be divided easily based on the grayscale value of various points. The segmentation of thresholds is further split into the local threshold and global threshold methods. The image is divided into two regions using the global threshold method through the single threshold. One region is used as the target region, and the other region is used as the background. In the local threshold method, several segmentation thresholds are required. So, the image is divided into numerous target regions and backgrounds [1]. Usually, the Ostu's method is employed in which optimization of variance between groups is done and a global optimal threshold is identified. Also, many other methods that likely come under threshold segmentation are the relaxation method, moment preserving technique, statistical mode, entropy-based technique, co-occurrence method, and fuzzy set method [22]. Different thresholding techniques are shown in Figure 8.1.

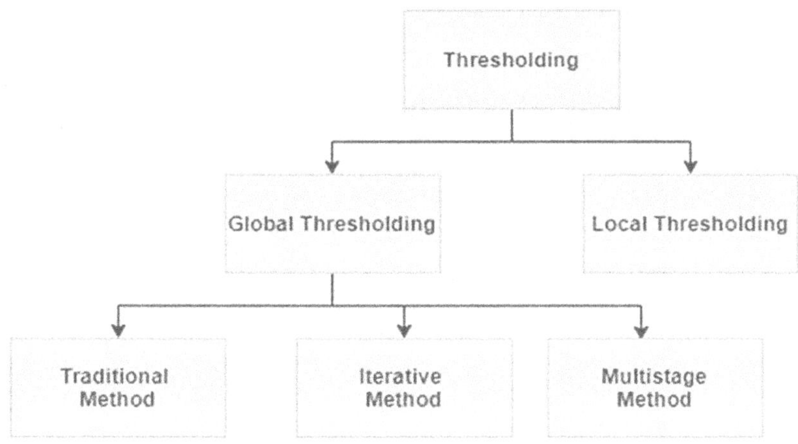

FIGURE 8.1 Thresholding techniques.

TABLE 8.2
Advantages and Disadvantages of Threshold Segmentation and Regional Growth Segmentation

Advantage	In the threshold method, the estimate analysis is easy and the speed of action is better, specifically, in the variation in the target and background results in the segmentation effect.
	The benefit of regional growth segmentation is its ability to divide the associated regions with identical features and to offer decent threshold data with the outcomes of segmentation. It clearly defines regional development and needs a limited seed point to achieve this. Also, the parameters for growth can be freely defined in the growth period. Finally, at the same time, it will choose several parameters.
Disadvantage	The main drawbacks of image segmentation are the lack of substantial grayscale variance, overlapping the values of grayscale in image. This causes difficulties in obtaining precise results as it only considers the image's gray information without taking into account the image's spatial information; it is vulnerable to noise and grayscale unevenness, and mixed with other approaches.
	The disadvantage of the regional growth segmentation is that the cost of computing is high. Noise and grayscale unevenness may also add to voids and separation. The last one is that the picture's shadow effect is always not very good.

8.2.1.2 Regional Growth Segmentation

This approach is the same as the serial region-based method, and its concept is to form a region of those pixels that have identical properties and features. In this method, the initial step is to select the seed pixels. In the next step, identical pixels are combined around the identical pixel in the seed pixel location region. Some of the advantages and disadvantages of both threshold segmentation and regional growth segmentation are mentioned in Table 8.2.

8.2.2 Edge Detection Segmentation

Edges are curves in which there are abrupt brightness shifts or spatial brightness derivatives. Brightness variations are where the orientation of the surface changes discontinuously, where one object obscures another, where shadow lines appear, or where the properties of the surface reflection are discontinuous. To maximize the visibility of distorted images, an edge-detection filter may also be used. Edge detection for image segmentation is one of the most significant applications. Image segmentation is used to execute the partition of a digital image in several regions or pixel sets [23]. The edge is the border of two homogeneous regions. The detection of edges refers to the method of recognizing and finding the sharp variance in a picture. Edge detection is used to identify artifacts that are useful for a variety of applications, including medical image processing, biometrics, and so on. As it enables higher-level image processing, edge detection is an active field of study [24]. Edge-detection methods based on discontinuity are the most widely used methods.

+1	+2	+1
0	0	0
-1	-2	-1

(a)

-1	0	+1
-2	0	+2
-1	0	+1

(b)

FIGURE 8.2 (a) Gx, (b) Gy.

Roberts edge detection, Sobel edge detection, Prewitt edge detection, Kirsh edge detection, Robinson edge detection, LoG edge detection, and Canny edge detection are these techniques.

8.2.2.1 Sobel Edge Operator

Using the Sobel approximation to the derivative, the Sobel method of edge detection for image segmentation seeks edges. It precedes the edges at those points where the gradient is strongest. The Sobel method computes a 2-D spatial gradient quantity on an image, highlighting high-spatial frequency regions that correspond to edges. It's commonly used to find the estimated absolute gradient magnitude in each point of an input grayscale image [24]. Using the discrete discrepancies between the horizontal and vertical lines of a 3×3 neighbors shown in Figure 8.2, the Sobel edge detector computes the gradient. The Sobel operator is based on a thin, independent, binary-valued filter converging the image. Figure 8.2 represents the process of the Sobel edge operator.

8.2.2.2 Roberts Edge Detector

This is employed for conducting simple, easy-to-compute 2-D spatial gradient measurement on an image. This approach focuses on the regions having large spatial variances that also summarize a given dataset. The operator input is the grayscale picture that is most commonly used for this procedure, the same as the output [24]. The maximum magnitude of the spatial gradient in the input image at that point is represented by the pixel values at each point in the output. Figure 8.3 shows Robert's convolution mask.

1	0
0	-1

0	+1
-1	0

FIGURE 8.3 Roberts's convolution mask.

8.2.2.3 Prewitt Edge Detector

This edge detector is a valid method for measuring the edge's magnitude and direction. However, separate gradient edge detection requires a very slow calculation for estimating the path using magnitudes in the x and y directions; compass edge detection, on the other hand, gets the path straight from the kernel with the best result. It has only eight possible directions; however, experience indicates that it is no longer suitable for other direct path measurements. The Prewitt method is the oldest and the best method for edge detection [25].

8.2.2.4 Canny Edge Detector

The Canny edge detection method is one of the basic techniques. The Canny method is a very significant way of identifying edges by separating noise from the image before finding the edges of the image. Without disrupting the characteristics of edges in the picture, the Canny approach is a safer way to apply the inclination to locate the edges and the value of the threshold.

8.2.3 Clustering Segmentation Method

A common theory related to image segmentation does not exist. Though numerous new concepts and approaches have been developed in other fields, certain methods of image segmentation have been merged with some particular algorithms. The supposed class has found its application for the set of identical elements. Clustering is commensurate with the standards and rules related to grouping of objects in the system [26]. To segment the pixels in the image space with the corresponding space points, the image space clustering technique is used. Some of the clustering methods are explained below:

a. **k-Means Clustering:** Based on the distance of pixels from each other, the k-means algorithm divides the data into groups of multiple clusters. k-means consider the medium level of computation and make a correspondingly high level of clusters [27]. To minimize the distance between the pixels and cluster center, k-means algorithm is used. Clustering the data means grouping similar objects together. Based on the similarity of pixels, the k-means paired the data into one group. The k-means algorithm is numerical, performs well with good data, and is a non-deterministic and unsupervised method [1]. The steps that are followed in the processing of k-means are shown in Figure 8.4. There are three basic limitations of the k-means algorithm.
   1. It is important to define the number of clusters.
   2. Various initial conditions yield various outcomes.
   3. The data far from the center pushes the center away from the optimal spot.

b. **Fuzzy c-Means (FCM):** FCM is one of the unofficial clustering algorithms that is applied to a wide variety of attribute analysis, clustering, and classification architecture concerns. Agricultural engineering, physics, chemistry, geology, image analysis, medical diagnosis, form analysis, and target recognition are only a few of the applications of FCM.

The fuzzy c-means clustering algorithm was proposed in the 1980s with the development of the fuzzy theory. Grouping of the same type of data into one cluster is done by fuzzy c-means clustering [28]. This clustering is accomplished iteratively minimizing a cost function that relies on the distance of the pixels to the cluster centers in the feature domain. The pixels on an image are strongly clustered, i.e., the pixels have approximately the same attribute details in the immediate neighborhood. The steps followed in the processing of fuzzy c-means are shown in Figure 8.5.

8.3 CONCLUSION

To solve various image processing problems related to many images, such as medical images, leaf disease images, vehicle images, etc., segmentation algorithms are used. They divide the image data into meaningful and segmented part that is understood by the user easily. With these methods, the segmentation of data is done very easily and that segmented data is used for further processing. Segmentation helps in classification phase to classify data according to data from the user.

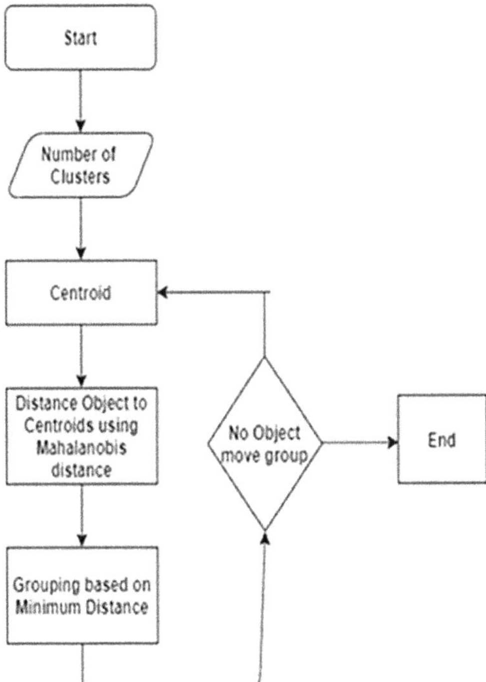

FIGURE 8.4 k-means clustering process.

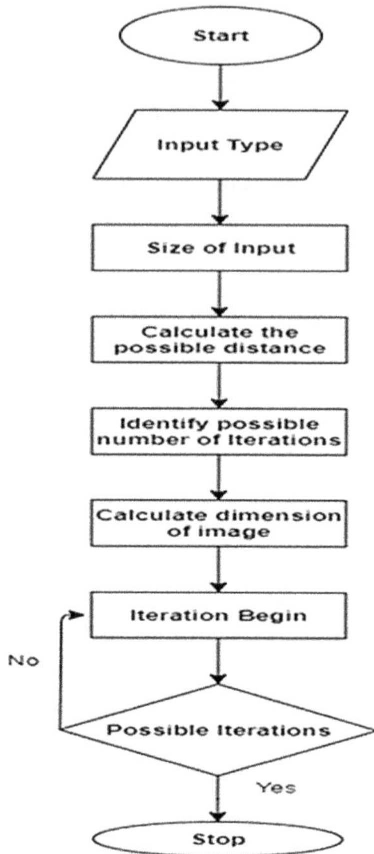

FIGURE 8.5 Fuzzy c-means clustering process.

REFERENCES

1. S. Yuheng and Y. Hao, "Image segmentation algorithms overview," *arXiv*, vol. 1, (2017), doi.org/10.48550/arXiv.1707.02051.
2. S. Minaee, Y. Y. Boykov, F. Porikli, A. Plaza, N. Kehtarnavaz, and D. Terzopoulos, "Image segmentation using deep learning: A survey," *IEEE PAMI*, (2021), doi: 10.1109/TPAMI.2021.3059968.
3. V. Kukreja and P. Dhiman, "A deep neural network based disease detection scheme for citrus fruits," *Proc. - Int. Conf. Smart Electron. Commun. ICOSEC 2020*, Icosec (2020), pp. 97–101, doi: 10.1109/ICOSEC49089.2020.9215359.
4. L. Antonelli, V. D. Simone, and D. Serafino, "A view of regularized approaches for image segmentation," *arXiv:2102.05533v3*, (2021), doi: 10.48550/arXiv.2102.05533.
5. A. Garcia-Garcia, S. Orts-Escolano, S. Oprea, V. Villena-Martinez, P. Martinez-Gonzalez, and J. Garcia-Rodriguez, "A survey on deep learning techniques for image and video semantic segmentation," *Appl. Soft Comput.*, vol. 70, (2018) pp. 41–65, doi: 10.1016/j.asoc.2018.05.018.

6. S. Kalaivani, S. P. Shantharajah, and T. Padma, "Agricultural leaf blight disease segmentation using indices based histogram intensity segmentation approach," *Multimed. Tools Appl.*, vol. 79, no. 13–14, (2020) pp. 9145–9159, doi: 10.1007/s11042-018-7126-7.
7. S. Zhang, H. Wang, W. Huang, and Z. You, "Plant diseased leaf segmentation and recognition by fusion of superpixel, K-means and PHOG," *Optik.*, vol. 157, (2018) pp. 866–872, doi: 10.1016/j.ijleo.2017.11.190.
8. M. Sharif, M. A. Khan, Z. Iqbal, M. F. Azam, M. I. U. Lali, and M. Y. Javed, "Detection and classification of citrus diseases in agriculture based on optimized weighted segmentation and feature selection," *Comput. Electron. Agric.*, vol. 150, (2018) pp. 220–234, doi: 10.1016/j.compag.2018.04.023.
9. Q-xia Hu, J. Tian, and D. J. He, "Wheat leaf lesion color image segmentation with improved multichannel selection based on the Chan–Vese model," *Comput. Electron. Agric.*, vol. 135, (2017) pp. 260–268, doi: 10.1016/j.compag.2017.01.016.
10. N. Valliammal and S. N. Geethalakshmi, "Leaf image segmentation based on the combination of wavelet transform and k means clustering," *Int. J. Adv. Res. Artif. Intell.*, vol. 1, no. 3, (2012) pp. 37–43, doi: 10.14569/ijarai.2012.010307.
11. J. G. A. Barbedo, "A novel algorithm for semi-automatic segmentation of plant leaf disease symptoms using digital image processing," *Trop. Plant Pathol.*, vol. 41, no. 4, (2016) pp. 210–224, doi: 10.1007/s40858-016-0090-8.
12. R. A. Naqvi, D. Hussain, and W. K. Loh, "Artificial intelligence-based semantic segmentation of ocular regions for biometrics and healthcare applications," *Comput. Mater. Contin.*, vol. 66, no. 1, (2021) pp. 715–732, doi: 10.32604/cmc.2020.013249.
13. U. Jamil, A. Sajid, M. Hussain, O. Aldabbas, A. Alam, and M. U. Shafiq, "Melanoma segmentation using bio-medical image analysis for smarter mobile healthcare," *J. Ambient Intell. Humaniz. Comput.*, vol. 10, no. 10, (2019) pp. 4099–4120, doi: 10.1007/s12652-019-01218-0.
14. J. Sun, Y. Peng, Y. Guo, and D. Li, "Segmentation of the multimodal brain tumor image used the multi-pathway architecture method based on 3D FCN," *Neurocomputing*, vol. 423, (2021) pp. 34–45, doi: 10.1016/j.neucom.2020.10.031.
15. I. Aranguren, A. Valdivia, B. Morales-Castaneda, D. Oliva, M. A. Elaziz, and M. Perez-Cisneros, "Improving the segmentation of magnetic resonance brain images using the LSHADE optimization algorithm," *Biomed. Signal Process. Control*, vol. 64, (2021), pp. 102259, doi: 10.1016/j.bspc.2020.102259.
16. J. G. A. Barbedo, "A new automatic method for disease symptom segmentation in digital photographs of plant leaves," *Eur. J. Plant Pathol.*, vol. 147, no. 2, (2017) pp. 349–364, doi: 10.1007/s10658-016-1007-6.
17. J. -H. Zhang, F. -T. Kong, J. Zhai Wu, S. Qing Han, and Z. Fen Zhai, "Automatic image segmentation method for cotton leaves with disease under natural environment," *J. Integr. Agric.*, vol. 17, no. 8, (2018) pp. 1800–1814, doi: 10.1016/S2095-3119(18)61915-X.
18. V. Singh, "Sunflower leaf diseases detection using image segmentation based on particle swarm optimization," *Artif. Intell. Agric.*, vol. 3, (2019) pp. 62–68, doi: 10.1016/j.aiia.2019.09.002.
19. J. P. Kumar and S. Domnic, "Image based leaf segmentation and counting in rosette plants," *Inf. Process. Agric.*, vol. 6, no. 2, (2019) pp. 233–246, doi: 10.1016/j.inpa.2018.09.005.
20. K. Tian, J. Li, J. Zeng, A. Evans, and L. Zhang, "Segmentation of tomato leaf images based on adaptive clustering number of K-means algorithm," *Comput. Electron. Agric.*, vol. 165, August (2019) p. 104962, doi: 10.1016/j.compag.2019.104962.
21. D. Riehle, D. Reiser, and H. W. Griepentrog, "Robust index-based semantic plant/background segmentation for RGB- images," *Comput. Electron. Agric.*, vol. 169, December (2020), p. 105201, doi: 10.1016/j.compag.2019.105201.

22. N. Senthilkumaran and S. Vaithegi, "Image segmentation by using thresholding techniques for medical images," *Comput. Sci. Eng. An Int. J.*, vol. 6, no. 1, (2016) pp. 1–13, doi: 10.5121/cseij.2016.6101.
23. S. S. Al-amri, N. V. Kalyankar, and S. D. Khamitkar, "Image segmentation by using edge detection," *Int. J. Comput. Sci. Eng.*, vol. 2, no. 3, (2010) pp. 804–807.
24. M. M. Radha, "Edge detection techniques for Image Segmentation," *Int. J. Comput. Sci. Inf. Technol.*, vol. 3, no. 6, (2011) pp. 259–267.
25. P. P. Acharjya, R. Das, and D. Ghoshal, "Study and comparison of different edge detectors for image segmentation," *Glob. J. Comput. Sci. Technol. Graph. Vis.*, vol. 12, no. 13, (2012) pp. 29–32.
26. N. Dhanachandra, K. Manglem, and Y. J. Chanu, "Image segmentation using k-means clustering algorithm and subtractive clustering algorithm," *Procedia Comput. Sci.*, vol. 54, (2015) pp. 764–771, doi: 10.1016/j.procs.2015.06.090.
27. P. Kaur and V. Gautam, "Plant biotic disease identification and classification based on leaf image: A review," *Proc. 3rd Int. Conf. Comput. Inform. Networks. Lecture Notes in Networks and Systems,* vol. 167. Springer, (2021) pp. 597–610, doi: 10.1007/978-981-15-9712-1_51.
28. A. M. Mishra and V. Gautam, "Weed species identification in different crops using precision weed management: A review," *CEUR Workshop Proc.*, vol. 2786, (2021) pp. 180–194.

9 Alzheimer's Disease Classification

Monika Sethi, Sachin Ahuja, and Vinay Kukreja
Chitkara University Institute of Engineering & Technology,
Chitkara University
Punjab, India

CONTENTS

9.1 INTRODUCTION

Alzheimer's disease (AD) is a severe chronic neurodegenerative impairment [1] that really has no known treatment [2] or a precise deterministic definition of the disease process itself. While in the later stages, indicators or the markers of the disease are amyloid plaques, neurofibrillary tangles, and neurological damage, it is just not known how they are initially formed. This disease was named after Dr. Alois Alzheimer [3]. He found significant differences in the brain tissue of a woman who died due to an abnormal psychological illness. Her symptoms included the loss of memory, speech impediments, and irrational behavior. When she expired, he analyzed her brain and identified numerous unusual clumps (now known as amyloid plaques) and clusters of fibers (now termed as tau tangles or neurofibrillary tangles). Such plaques and tangles in the brain are still believed to be among the main symptoms of AD. Yet another feature is the deterioration of connections among neurons in the brain.

This causes the hippocampus and cortex to shrink and the cerebral ventricles to widen. The severity of these disruptions depends on the stage of the disease. In the later phases or stages of AD, the substantial shrinkage of the hippocampus [4] and cerebral cortex and the expansion of the ventricles can be seen clearly in the brain scans (medical resonance images or MRI) [5, 6]. Sufferers at the early stages of AD have Mild Cognitive Impairment (MCI), but not all sufferers with MCI necessarily develop AD. MCI is the initiation phase between the mental changes that are seen in normal aging and early-stage of AD, in which the individual endures minor behavior changes that are apparent to just the individual and their loved ones only [7].

DOI: 10.1201/9781003143468-9

MCI sufferers then are considered MCI converters (MCIc) or MCI non-converters (MCInc), which suggests that they have or have not progressed to AD during a period of one and a half years. Subsequently, there are two subtypes of MCI that are seldom described in the literature: early MCI (eMCI) and late MCI (lMCI) [8].

Approximately 5.7 million people are affected by AD. The number of AD sufferers is expected to increase by 13.8 million by the mid-century. More than 16 million involuntary nurses and relatives of the patient delivered nearly 18.4 billion hours of treatment to AD patients. The gross expense was more than $232 billion [9].

Diagnosing AD is a huge challenge, particularly in the initial stages. Previous noninvasive assessment methods imposed a strong focus on individual medical history, cognitive tests, and behavioral assessments conducted by healthcare professionals. Research findings also show that the language ability of most AD patients is affected, so neuropsychological tests are typically an effective way to detect AD at initial stages. These tests use a variety of measures that include a set of questions related to attention, language, orientation, and visual spatial skills. For instance, Mini Mental State Assessment (MMSA) is a standard cognitive psychological test. It requires the patient to respond to a variety of questions and after that, the doctor assigns a score between zero and thirty representing various cognitive skills. The validity of this method completely depends on the level of expertise and knowledge of the healthcare professional. Given the exponential spike in the number of patients, this type of test requires a lot of time and resources.

Researchers have recently established the susceptibility of various biomarkers for earlier diagnosis of both AD and MCI [10]. To identify abnormal pathological volumetric changes associated with AD, biomarkers from structural MRI can be used to measure brain atrophy. Functional imaging like positron emission tomography (PET) scans can also provide complementary information related to the hypo-metabolism quantification, while cerebrospinal fluid gives evidence of protein changes.

Machine Learning (ML) techniques have been shown to be very effective for diagnosing AD. The classification system consists of multiple phases, such as extraction of features from the given data, selection of features, mapping the features into lower dimensions, and then classification based on those selected features. Generally, to build any predictive model to facilitate the automation, decision support in the medical fields that minimally features extraction and classification algorithm is needed. However, Deep Learning (DL) incorporates the feature extraction phase into the learning model itself. For large datasets, DL is found to be especially useful for image data. Shi et al. indicate that a blend of neural networks (NN) and intelligent agents can be useful for analyzing medical images. In this chapter, various ML and DL techniques are explored related to their performance in classification of AD.

9.2 DATASETS FOR AD DIAGNOSIS

Various open access AD dataset repositories are available to assist researchers in accelerating their research in this area. Datasets such as Alzheimer's Disease Neuroimaging Initiative (ADNI) (http://adni.loni.usc.edu/), Open Access Series of Imaging Studies (OASIS) (www.oasis-brains.org), Minimal Interval Resonance

Imaging in Alzheimer's Disease (MIRIAD) (http://miriad.drc.ion.ucl.ac.uk/), and Australian Imaging, Biomarkers & Lifestyle Flagship Study of Ageing (AIBL) (aibl.csiro.au) are commonly used in this area of research [11]. The studies aim to test the feasibility of using various brain scans as an outcome measure for clinical trials of AD treatments, and these datasets generally include various types of neuroimaging modalities, such as MRI, functional MRI (fMRI), PET, Single-Photon Emission Computerized Tomography (SPECT), cognitive and clinical assessments, and the demographic information of patients.

Among all the datasets stated before, ADNI is being extensively used alone and with a blend of non-ANDI by a few researchers. ADNI was released by commercial healthcare organizations, including the National Institute of Aging (NIA), National Institute of Basic Biology (NIBB), and U.S. Food & Drug Administration (FDA), 2003. The North American-based study was designed for 800 adult subjects in total (400 individuals with MCI and 200 each for AD and Normal Controls (NC)). These subjects/groups were tracked by concerned organization for a couple of years. Subjects were hired from more than 50 different locations across Canada and the United States. The ultimate goal of the ADNI is to examine whether the neurological scans (such as MRI, fMRI, PET, etc.), genetics, and psychiatric methods can be incorporated to evaluate the various stages of AD.

The next dataset, MIRIAD, comprises 69 longitudinal volumetric MRI brain scans in total (46 AD subjects and 23 NC subjects). 46 AD subjects and 23 NC subjects were followed by the organization from 2 weeks to 2 years, with multiple scans. Lastly, it comprises around 700 scans acquired by the same scanner as well as the same radiologist. Additional information, such as gender (male or female), age and MMSE points, were also considered.

OASIS (www.oasis-brains.org) is an initiative that offers free datasets to the academic world, comprising longitudinal and cross-sectional MRI brain scans for 150 subjects and 416 subjects, respectively. The longitudinal dataset includes the neuroscans for older people (from 60 to 96 years old) while the cross-sectional dataset comprises the scans of young people to elderly people (from 18 to 96 years old). A series of studies employed different datasets for training and test purposes; for instance, in [3], the researchers used the OASIS dataset for training and the MIRIAD dataset for test purposes to train a DL Convolutional Neural Network (CNN) model for AD classification.

Another dataset, AIBL, which was released in 2006, aims at exploring the biomarkers involving cognitive, health, and lifestyle characteristics that are useful for identifying the distinct stages of AD. It entails more than 1,000 participants, including different groups, such as AD, MCI, and NC.

To preprocess the data, past researchers employed software such as Statistical Parametric Mapping (SPM) and FreeSurfer Library (FSL). FSL helps for brain extraction and tissue segmentation, whereas SPM helps align, spatially normalize, and smooth the brain scans. FreeSurfer has a pre-processing stream that involves three things, namely, skull stripping, non-linear registration, and segmentation. In [12], the researcher groups utilized a Clinica software framework proposed by ARAMIS Lab that facilitates the working of FSL, SPM, and FreeSurfer.

9.3 AD DIAGNOSIS USING ML TECHNIQUES

Generally, neuroimaging data for efficient recognition and classification makes use of machine learning techniques such as Support Vector Machine (SVM), K-nearest neighbor (KNN), Random Forest, Naive Bayes, etc. These are based on solitary classifiers, and SVM is the most well-known of these methods [13]. This is a very stable and commonly used technique for regression [14] as well as classification [15] problems. By including the Systemic Risk Minimization Theory (SRM), SVM offers strong generalization efficiency. To define the data points, SVM uses the maximal margin principal. Classification accuracy of SVM is heavily affected by mainly two parameters, namely, the selection of the features and the setting of the kernel parameter. To classify AD, various kernel functions have been utilized for SVM [16] to translate the data to higher dimensions. Several researchers used the SVM linear kernel to identify AD results. This is because there is no kernel parameter to tune in the linear kernel. Some researchers have also used multiple kernels with SVM [10]. Table 9.1 shows the findings of various researchers for AD classification using the SVM linear kernel.

To extricate the highly representatives features from the mean neuro-images specific to AD patterns, Vicente et al. in their paper [17], proposed a novel CAD framework based on independent component analysis (ICA). The main purpose of ICA was to locate the independent component sources that are present during the AD phase. Furthermore, it was demonstrated to be a successful technique for feature reduction dimensionality and choosing the most pertinent information. Then, SVM was utilized to deal with the classification-related task. An accuracy of 87% in distinguishing AD from HC, with a sensitivity of 90% and specificity of 84%, was achieved.

Voxel-based Morphometry (VBM) from regular T1-weighted MRI neuroimages demonstrated effective results for measuring AD-related brain atrophy and to empower a genuinely precise classification of AD, MCI, and Healthy Controls (HC) patients. In [18], the authors evaluated two different VBM algorithms (in-house, they were named MorphoBox and FreeSurfer) on the ADNI dataset to classify AD automatically. Their results clearly indicated that both algorithms achieved accuracy equivalent to the traditional whole-brain VBM pipeline utilizing the SPM tool for AD vs. HC and MCI vs. HC and early vs. late AD converters, subsequently, exhibiting the capability of VBM to aid in the diagnosis of MCI and AD. The highest sensitivity and specificity was achieved for AD vs. HC and was 86% and 91%, respectively.

Authors in [19] utilized Sparse Inverse Covariance Estimation (SICE) to acquire undirected graphs for connectivity patterns from multimodalities MRI and PET neuro images (ADNI dataset). They included three different subjects for their research, namely HC (68), AD (70), and MCI (111). AD vs. HC achieved 8% more accuracy as compared to AD vs. MCI.

The SVM-Recursive Feature Elimination (RFE) method was implemented in [4, 20] to classify AD, cMCI, ncMCI, and HC subjects of sMRI neuro images. The dataset was drawn from ADNI. The maximum accuracy (89%) was achieved while

TABLE 9.1
AD Classification Using SVM Linear Kernel

S. No.	Ref. No. (Year)	Dataset	Neuroimaging Modality (Single or Multimodality)	Sample Size	Feature Extraction	Binary Classification	Performance Metrics (%age) SEN	SPE	ACC
1	(2015) [17]	ADNI	sMRI (Single)	AD (188), HC (229), MCI (401)	VBM and FastICA	HC/AD	90	84	87
						HC MCI	80	74	78
						AD/MCI	86	85	85
2	(2015) [18]	ADNI	sMRI (Single)	AD (188), HC (229), MCI (401)	VBM and VoIBM	HC/AD	86	91	NM
						HC/MCI	78	68	
						AD/MCI	69	67	
3	(2015) [19]	ADNI	sMRI+PET (Multimodality)	AD (70), HC (68), MCI (111)	VCM and SICE	HC/AD	96	86	92
						HC/ MCI	90	82	86
						AD/MCI	87	81	84
4	(2015) [20]	ADNI	sMRI (Single)	AD (144), HC (189), cMCI (136), ncMCI (166)	VB and SNM-RFE	HC/AD			89
						cMCI/ ncMCI	NM	NM	71
5	(2017) [21]	ADNI	sMRI (Single)	AD(160), HC(162), sMCI(65), pMCI (71)	VBM+GA	HC/AD	89	97	93
						pMCI/sMCI	77	73	75
6	(2017) [22]	ADNI	sMRI (Single)	AD (65), HC (135), cMCI (132), pMCI (95)	MDS+PCA	HC/AD	93	98	97
						HC/pMCI	87	95	92
						sMCI/pMCI	86	91	89
7	(2018) [23]	ADNI	DTI (Single)	AD(48), HC(51), eMCI(75), lMCI (39)	LDH and SVM-RFE	HC/AD	NM	NM	90
						HC/eMCI			88
						HC/lMCI			100
						eMCI/lMCI			93
						eMCI/AD			85
						lMCI/AD			98

(Continued)

TABLE 9.1 *(Continued)*

AD Classification Using SVM Linear Kernel

S. No.	Ref. No. (Year)	Dataset	Neuroimaging Modality (Single or Multimodality)	Sample Size	Feature Extraction	Binary Classification	Performance Metrics (%age) SEN	SPE	ACC
8	(2019) [24]	ADNI	fMRI (Single)	AD(24), HC(24), eMCI(24), lMCI (24)	RF-score	HC/AD	NM	NM	96
						HC/MCI			94
						HC/lMCI			96
						eMCI/lMCI			88
						lMCI/AD			92
9	(2020) [25]	ADNI	fMRI (Single)	AD(50), HC(50), MCI (50)	USVM- RFE	HC/AD	NM	NM	100
						HC/MCI			90
						MCI/AD			74

Abbreviations: **ADNI**, Alzheimer's Disease Neuroimaging Initiative; **HC**, Healthy Control; **AD**, Alzheimer's disease; **MCI**, Mild Cognitive Impairment; **eMCI**, Early MCI; **lMCI**, Late MCI; **sMCI**, Stable MCI; **pMCI**, Progressive MCI; **fMRI**, Functional Magnetic Resonance Imaging; **NM**, Not Mentioned classifying AD with HC against the other binary classification, cMCI vs. ncMCI (71%).

In the paper [21], the research group proposed an automatic feature-selection method (FSM) for AD classification as well as for predicting the conversion time from MCI to AD, reliant on a genetic algorithm (GA). The FSM was realized by extricating the voxel-values as a raw feature from the volume of interest (VOI) attained from VBM analysis. The extricated raw features vectors were then lessened to lower-dimensional feature vectors. Using sMRI neuro images drawn from the ADNI dataset, the system achieved an accuracy of 93% and 75% for AD/HC and sMCI/pMCI, respectively.

In [22], the authors designed an ML model to classify AD or MCI subjects from HC and to predict the MCI to AD conversion time by evaluating and examining the regional morphological changes of neuro-scans. sMRI neuroimaging modality data samples (AD-65, HC-135, cMCI- 132, and pMCI-95) were acquired from the ADNI dataset. Their technique achieved an accuracy of 97% in classifying mild AD patients from HC subjects considering the whole brain Gray Matter (GM) or temporal lobe (TL) as the region of interest (ROI), 92% in differentiating pMCI from HC, and 89% in classifying pMCI with sMCI using the hippocampus and amygdale as ROI.

The authors of [23] applied the ML technique to identify the features related to AD and MCI using Diffusion Tensor Imaging (DTI) neuro images (AD: 48 subjects, lMCI: 39 subjects, eMCI: 75 subjects, and HC: 51 subjects) drawn from ADNI dataset. SVM-RFE and Logistic Regression (LR) were combined using cross validation leave-one-out in order to identify features optimally. The outcome of the model showed that the SVM classifier performed better in terms of stability than the LR classifier. Their findings revealed a suggestion for ML-based image analysis on clinical diagnosis. The accuracy level achieved for MCI vs. HC subjects was 100%.

9.4 AD DIAGNOSIS USING DL

ML techniques are not adequate for dealing with complex problems such as AD classification. The increased computing power of Graphics Processing Units (GPU) has allowed the development of cutting edge DL algorithms [26]. DL is the branch of ML in artificial intelligence that mimics human intelligence in data analysis and pattern recognition to solve complex decision-making problems. DL approaches have revolutionized performance in a variety of areas, such as object detection, image segmentation, identification, audio processing, and mapping.

On the basis of neuroimaging scans, DL models can uncover hidden artifacts, find linkages between various sections of images, and recognize disease-related features. DL models have been successfully extended to medical images such as structural and functional MRIs, DTI, and PET. In this way, researchers have recently started to use DL models to detect medical images.

As the most extensively used DL design, CNN has obtained a considerable attention on its popularity [27] in computer vision applications such as image analysis, speech recognition, natural language processing, and most recently in the field of medical image classification and prediction. Its design is influenced by the concept of natural perceptual cortical proposed by Hubel and Wiesel in 1959 [28]. It is based on the theory of receptive areas, i.e., the region of the source images. In 1980, following encouragement from the research of Hubel and Weisel, Fukushima [29] developed a recognition that could be considered the basis of CNN architecture. Further, in 1990 LeCun et al. [30] established the structure of CNN by designing a multi-layer artificial model, that was referred to as LeNet-5, and seems to distinguish handwritten digits. Additionally, other Neural Networks (NNs) could be trained with a back propagation algorithm that provided an opportunity to retrieve different patterns directly from raw images, thereby combating the pre-processing steps needed for the extraction of features. Table 9.2 shows the findings of various researchers for AD classification using CNN.

Authors of [31] designed and tested a predictive classification model that combined both a sparse auto-encoder (SAE) and CNNs (2D CNN and 3D CNN) to distinguish AD, MCI from HC subjects using MRI images of the ADNI dataset [31]. SAE has been utilized to train the filters for the first convolution layer of two different CNN architectures (2D CNN and 3D CNN). Local patches of MRI images were used as an input. The convolution activities were trailed by pooling tasks, like in the neural system. Then the pooling layers were preceded by a fully connected layer and a softmax layer with probabilities of three independent subjects (AD, MCI, and HC) as output. Experimental results show that the accuracy for the binary classification of AD/HC in both 2D and 3D CNNs is same, i.e., 95.39%. There is a minor improvement in accuracy for AD/MCI, MCI/HC, and AD/MCI vs. HC with an increment of 4.6%, 1.98%, and 3.94%, respectively, for 3D CNN over 2D CNN. Although the authors have achieved the best results with their designed model, which uses the combination of SAE and 3D CNN, the accuracy can be improved further by carrying out in-depth searches for the best hyper-parameters in both architectures. Furthermore, the overall accuracy in the future can be enhanced using fine-tuning at the cost of increased computational power during training time.

In another research paper, Saraf et al. designed the CNN model for the classification of healthy subjects and brains affected by AD in adults (>75 years) using MRI and fMRI images of the ADNI dataset [5]. The reported accuracy was 99.9% for fMRI and 98.84% for MRI images.

In [32, 33] studies, the researchers performed a linear registration, and ROIs were extracted using an automated anatomical atlas (AAL). These ROIs are defined as 3D bounding boxes covering all the voxels of the hippocampus. A "2D+ε approach" was used in these three studies, i.e., three neighboring 2D slices in the hippocampus region of MRI using the ADNI dataset were used to make a patch. Per patient, only one or three patches were used. So, they did not cover the whole brain region. In the first study [32], one patch comprising only sagittal view slices was seen as an input to the CNN classifier. And then, the CNN comprised two CL layers followed by max-pooling and one top-level fully connected layer for AD classification. In the second study [40], the authors considered all three projections (axial, sagittal, and coronal) of the hippocampus region of the brain and then generated three patches per

TABLE 9.2
AD Classification Using CNN

S. No.	Ref. No. (Year)	Dataset	Neuroimaging Modality (Single or Multimodality)	Sample Size	Data Handling Technique	DL Model	Binary Classification	ACC (%age)
1	(2015) [31]	ADNI	sMRI (Single)	AD (755) MCI (755) HC (755)	Slice-based	Sparse Encoder with 2D CNN	AD/HC	95
							AD/MCI	82
							MCI/HC	90
					Voxel-based	Sparse Encoder with 3D CNN	AD/HC	99
							AD/MCI	87
							MCI/HC	92
2	(2016) [5]	ADNI	sMRI	AD (211) HC (91)	Slice-based	2D CNN	AD/HC	98.8
			rs-fMRI	AD (52) HC (92)				99.9
3	(2017) [32]	ADNI	sMRI	AD (188) MCI (399) HC(288)	Slice-based	2D CNN	AD/HC	82
							AD/MCI	62
							MCI/HC	60
4	(2017) [33]	ADNI	sMRI + PET (Multimodality)	AD (145) MCI (192) HC (172)	Patch-based (Patch size=3)	SAE+ 3D CNN	AD/HC	93
							AD/MCI	82
							MCI/HC	89
					(Patch size=5)		AD/HC	94
							AD/MCI	83
							MCI/HC	93
					(Patch size=7)		AD/HC	91
							AD/MCI	84
							MCI/HC	91

(Continued)

TABLE 9.2 (*Continued*)
AD Classification Using CNN

S. No.	Ref. No. (Year)	Dataset	Neuroimaging Modality (Single or Multimodality)	Sample Size	Data Handling Technique	DL Model	Binary Classification	ACC (%age)
5	(2017) [34]	ADNI	sMRI	AD (47) HC (34)	Slice-based	2D CNN	AD/HC	93
6	(2018) [35]	ADNI	sMRI +DTI	AD (48) MCI (108) HC(58)	ROI-based	3D CNN	AD/ HC	97
							AD/MCI	80
							MCI/HC	66
7	(2018) [36]	ADNI	sMRI	AD(150) HC(391)	Slice-based	2D CNN	AD/HC	95.9
						3D CNN	AD/HC	96.8
8	(2019) [27]	ADNI	sMRI	AD (647) sMCI (441) pMCI (326) HC (731)	ROI-based	3D CNN	AD/HC	81.19
							HC/pMCI	-
			PET				AD/HC	89.11
							HC/pMCI	-
			MRI+PET				AD/HC	90
							sMCI/pMCI	87.46
9	(2019) [37]	ADNI	sMRI+PET	AD (93) HC(100)	ROI-based	3D CNN	AD/HC	95
10	(2021) [38]	Kaggle Dataset	MRI	Non-demented (510) Mild demented (46) Very mild demented (50) Moderately demented (19)	-	CNN (Alexnet)	Precision Non-demented (94%) Mild demented (98%) Very mild demented (90%) Moderate demented (100%)	

(*Continued*)

TABLE 9.2 (*Continued*)
AD Classification Using CNN

S. No.	Ref. No. (Year)	Dataset	Neuroimaging Modality (Single or Multimodality)	Sample Size	Data Handling Technique	DL Model	Binary Classification	ACC (%age)
11	2021 [39]	OASIS-3	MRI	AD(170), NC(70),	Slice-based	2D CNN	AD/HC	99.3
12	(2019) [2]	ADNI	sMRI	AD (50) MCI (50) HC(50)	Slice-based	2D CNN	AD/HC AD/MCI HC/MCI	99 99.3 99.22
13	(2019) [13]	ADNI (A)	sMRI	AD (294) MCI (763) HC(352)	Slice-based	3D CNN	AD/HC (using dataset A)	99
		Non-ADNI (B)		AD(124) MCI(50) HC(55)			AD/HC (Using dataset A & B)	98

Abbreviations: **ADNI**, Alzheimer's Disease Neuroimaging Initiative; **HC**, Healthy Control; **AD**, Alzheimer's disease; **MCI**, Mild Cognitive Impairment; **ROI,** Region of Interest**; PET,** Positron Emission Tomography; **rs-fMRI**, Resting State Functional Magnetic Resonance Imaging subject: three separate CNNs comprising two CL, two pooling layers, followed by the ReLu activation function, and an FC layer. These networks were fused using two different techniques: intermediate and late fusion techniques. The reached accuracy was 91.4%, 69.5%, and 65.6% for AD vs. NC, AD vs. MCI and MCI vs. NC, respectively.

The authors presented a computer-aided AD recognition model, which was based on 3D CNN [34]. The brain's 3D topology was considered as a whole while AD classification. The CNN comprised three groups of processing layers, two fully connected layers, and one classifier layer. Each of the processing layers was made up of three layers, namely, convolutional layers, pooling layers, and normalization layer. Their model was trained and tested on the MRI images of the ADNI dataset.

The authors designed 3D CNN to combine the features from the hippocampus area of both the modalities T1-weighted MR and FDG-PET to classify AD [27]. In the case of the MRI, only the hippocampus area was chosen to be the ROI, but for PET images, the authors tried different ROIs that contained only the hippocampus area and both the hippocampi and cortices. They obtained the highest level of accuracy as 90.1%, 87.46%, and 76.9% for CN vs. AD, CN vs. pMCI, and sMCI vs. pMCI, respectively. These results demonstrated that using CNN segmentation is not the prerequisite. Although they achieved 90% accuracy for AD vs. CN classification, only 77% accuracy was recorded for classifying pMCI vs. sMCI. The performance of the model in terms of accuracy could be further enhanced by using new imaging modalities, such as T2-MRI, amyloid protein imaging.

To learn MRI features automatically in a target-oriented manner, various DL models have recently been propounded. However, feeding the whole MRI neuroimages to the CNN model directly did not generate the vigorous model, as millions of voxels are present in MRI and every part of the brain region may not be affected by AD. So, the main challenge in MRI-based DL models is identifying informative areas or regions in the neuroimaging modalities. To address this problem, the authors in [35] pre-extracted only the hippocampal ROI and neighboring areas in neuro scans (sMRI and DTI neuro images) from the ADNI dataset to train a CNN. The authors also demonstrated the size of the ROI matters for classification accuracy. The recorded accuracy was highest while classifying AD vs. HC (97%) and lowest for MCI vs. HC (66%). The authors in [36] propounded a 3D fine-tuned CNN model for classifying three brain subjects named HC, AD, and MCI, and the model was tested using the MRI ADNI dataset. They achieved accuracy for three binary classifications for AD/HC, MCI/HC, and AD/MCI as 97%, 93%, and 88%, respectively.

Due to inadequate medical images, it is rather hard to fully utilize CNN effectively. To overcome this issue, the authors in [37] designed a novel DL framework; particularly the goodness of 3D CNN and FSBi-LSTM was exploited. At first, they designed 3D CNN to extract the deep features from both the MRI and PET modalities using the ADNI dataset. To improve its performance further, the features maps output coming from 3D CNN was applied to FSBi-LSTM instead of the conventional FC layer to get rich semantic and spatial information. At last, they validated their model on the ADNI dataset. Binary classification accuracy reached up levels of 94.82%, 65.35%, and 86.36% for AD vs. NC, sMCI vs. NC, and pMCI vs. NC, respectively.

Various studies utilized DL models on MRI brain scans and achieved promising results. But most of the challenging issues with DL architectures such as CNN required a huge amount of dataset to train the model. The authors designed a Political, Economic, Social, Technological, Legal and Environmental factors (PESECTL) mathematical model based on transfer learning [2]. In their model, a 2D CNN-based VGG-16 model trained on the ImageNet dataset was utilized for feature extraction to classify ADNI sMRI slices among three different classes, namely AD, CN, and MCI. For every single subject (AD, CN, and MCI), the authors considered 32 slices based on the information entropy to prepare the dataset for their designed model. The three-way classification (CN vs. AD vs. MCI) accuracy reached 95.73% on the validation dataset. For future examinations, specialists should attempt other neural systems; for example, residual network, inception network, and later cutting edge systems as a base model for building the classifier.

Furthermore, the authors designed a DL algorithm based on the CNN architecture to discriminate AD, sMCI (MCI patients who remain stable and have not converted to AD), and cMCI (MCI patients who do not remain stable and have converted to AD) on 3D T1-weighted sMRI from combined ADNI and non-ADNI datasets. The CNN model was trained, validated, or tested on the ADNI dataset and combined ADNI and non-ADNI datasets separately to measure differences in accuracy [13]. Their model achieved almost similar accuracy up to 75% with only ADNI and combined ADNI+ non-ADNI dataset while discriminating cMCI and sMCI subjects. They excluded the presence of cMCI among s-MCI subjects. For this, a proper clinical follow-up is required. Their model should be tested with a combination of other biomarkers, such as PET, cognitive, genetic, and cerebrospinal fluid.

9.5 OBSERVATIONS

It has been observed that dataset selection is crucial and can influence the output of the classifier. The findings cannot be compared, even though studies used the same dataset, had the same number of subjects, and used the same subject number code. But still a few common observations, such as structural and functional neuroimaging modalities like MRI, fMRI, PET, and DTI, are very useful in identifying AD. Many researchers also use other parameters in addition to neuroimages such as age, educational level, cognitive-test points, and genetics to improve the accuracy of AD classification. It is still ambiguous to declare the most discriminatory neuroimaging modality. However, multimodalities (a combination of one or more modalities together) are considered to be most successful since they represent the complementary information and provide better results over single modality. Next, the classification for MCI vs. HC and the transition time from MCI to AD are more valuable than the classification for AD vs. HC. ROI and patch-based data handling strategies are more efficacious than slices and voxel-based techniques, as ROI-based methods lead to low dimension features and are readily understood.

9.6 CONCLUSION AND FUTURE DIRECTIONS

This chapter began with a description of AD and its symptoms, accompanied by an overview of the existing diagnostic criteria and associated biomarkers such as MRI, PET, and fMRI. In this chapter, papers using ML techniques, particularly SVM and DL model (CNN), were studied for AD diagnosis. As per this chapter, it can be noted that the most widely used dataset for AD is obtained from ADNI after gaining authorization from the concerned authority. More than 95% of researchers that we considered in this chapter have utilized the ADNI dataset. In both the ML and DL techniques, combining the modalities to identify AD gives better results. The advantage of using DL techniques over ML is the elimination of the feature extraction phase, which makes the work of researchers very fast and easy. The classification accuracy for AD vs. HC is always highest when compared with other binary classifications, such as AD vs. MCI and MCI vs. HC. So, in future, more work can be done to achieve comparable accuracy for cMCI vs. ncMCI, MCI vs. HC, and AD vs. HC, and to find the conversion time from the MCI stage to AD.

REFERENCES

1. Liu, M., Li, F., Yan, H., Wang, K., Ma, Y., Shen, L., & Alzheimer's Disease Neuroimaging Initiative. (2020). A multi-model deep convolutional neural network for automatic hippocampus segmentation and classification in Alzheimer's disease. *Neuroimage, 208*, 116459.
2. Jain, R., Jain, N., Aggarwal, A., & Hemanth, D. J. (2019). Convolutional neural network based Alzheimer's disease classification from magnetic resonance brain images. *Cognitive Systems Research, 57*, 147–159.
3. YİĞİT, A., & IŞIK, Z. (2020). Applying deep learning models to structural MRI for stage prediction of Alzheimer's disease. *Turkish Journal of Electrical Engineering & Computer Sciences, 28*(1), 196–210.
4. Jabason, E., Ahmad, M. O., & Swamy, M. N. S. (2019, August). Classification of Alzheimer's disease from MRI data using an ensemble of hybrid deep convolutional neural networks. In *2019 IEEE 62nd International Midwest Symposium on Circuits and Systems (MWSCAS)* (pp. 481–484). IEEE.
5. Sarraf, S., DeSouza, D. D., Anderson, J., & Tofighi, G. (2017). DeepAD: Alzheimer's disease classification via deep convolutional neural networks using MRI and fMRI. *BioRxiv*, 070441.
6. Kumar, S. S., & Nandhini, M. (2017). A comprehensive survey: Early detection of Alzheimer's disease using different techniques and approaches. *International Journal of Computer Engineering & Technology, 8*(4), 31–44.
7. Choi, B. K., Madusanka, N., Choi, H. K., So, J. H., Kim, C. H., Park, H. G., ... & Prakash, D. (2020). Convolutional neural network-based MRI image analysis for Alzheimer's disease classification. *Current Medical Imaging, 16*(1), 27–35.
8. Ebrahimighahnavieh, M. A., Luo, S., & Chiong, R. (2020). Deep learning to detect Alzheimer's disease from neuroimaging: A systematic literature review. *Computer methods and programs in biomedicine, 187*, 105242.
9. Lyu, G. (2018, October). A review of Alzheimer's disease classification using neuropsychological data and machine learning. In *2018 11th International Congress on Image and Signal Processing, BioMedical Engineering and Informatics (CISP-BMEI)* (pp. 1–5). IEEE.

10. Alam, S., Kwon, G. R., & Alzheimer's Disease Neuroimaging Initiative. (2017). Alzheimer disease classification using KPCA, LDA, and multi-kernel learning SVM. *International Journal of Imaging Systems and Technology*, *27*(2), 133–143.
11. Tanveer, M., Richhariya, B., Khan, R. U., Rashid, A. H., Khanna, P., Prasad, M., & Lin, C. T. (2020). Machine learning techniques for the diagnosis of Alzheimer's disease: A review. *ACM Transactions on Multimedia Computing, Communications, and Applications (TOMM)*, *16*(1s), 1–35.
12. Liu, S., Yadav, C., Fernandez-Granda, C., & Razavian, N. (2020, April). On the design of convolutional neural networks for automatic detection of Alzheimer's disease. In *Machine Learning for Health Workshop* (pp. 184–201). PMLR.
13. Basaia, S., Agosta, F., Wagner, L., Canu, E., Magnani, G., Santangelo, R., ... & Alzheimer's Disease Neuroimaging Initiative. (2019). Automated classification of Alzheimer's disease and mild cognitive impairment using a single MRI and deep neural networks. *NeuroImage: Clinical*, *21*, 101645.
14. García-Floriano, A., López-Martín, C., Yáñez-Márquez, C., & Abran, A. (2018). Support vector regression for predicting software enhancement effort. *Information and Software Technology*, *97*, 99–109.
15. Richhariya, B., & Tanveer, M. (2018). EEG signal classification using universum support vector machine. *Expert Systems with Applications*, *106*, 169–182.
16. Cristianini, N., & Scholkopf, B. (2002). Support vector machines and kernel methods: The new generation of learning machines. *Ai Magazine*, *23*(3), 31–31.
17. Vicente, J. M. F., Álvarez-Sánchez, J. R., De la Paz López, F., Toledo-Moreo, F. J., & Adeli, H. (Eds.). (2015). *Artificial Computation in Biology and Medicine: International Work-Conference on the Interplay between Natural and Artificial Computation, IWINAC 2015, Elche, Spain, June 1–5, 2015, Proceedings, Part I* (Vol. 9107). Springer. *Lect. Notes Comput. Sci. (including Subser. Lect. Notes Artif. Intell. Lect. Notes Bioinformatics)*, vol. 9107, pp. 78–87, 2015.
18. Schmitter, D., Roche, A., Maréchal, B., Ribes, D., Abdulkadir, A., Bach-Cuadra, M., ... & Alzheimer's Disease Neuroimaging Initiative. (2015). An evaluation of volume-based morphometry for prediction of mild cognitive impairment and Alzheimer's disease. *NeuroImage: Clinical*, *7*, 7–17.
19. Ortiz, A., Munilla, J., Álvarez-Illán, I., Górriz, J. M., Ramírez, J., & Alzheimer's Disease Neuroimaging Initiative. (2015). Exploratory graphical models of functional and structural connectivity patterns for Alzheimer's disease diagnosis. *Frontiers in Computational Neuroscience*, *9*, 132.
20. Retico, A., Bosco, P., Cerello, P., Fiorina, E., Chincarini, A., & Fantacci, M. E. (2015). Predictive models based on support vector machines: Whole-brain versus regional analysis of structural MRI in the Alzheimer's disease. *Journal of Neuroimaging*, *25*(4), 552–563.
21. Beheshti, I., Demirel, H., Matsuda, H., & Alzheimer's Disease Neuroimaging Initiative. (2017). Classification of Alzheimer's disease and prediction of mild cognitive impairment-to-Alzheimer's conversion from structural magnetic resource imaging using feature ranking and a genetic algorithm. *Computers in Biology and Medicine*, *83*, 109–119.
22. Long, X., Chen, L., Jiang, C., Zhang, L., & Alzheimer's Disease Neuroimaging Initiative. (2017). Prediction and classification of Alzheimer disease based on quantification of MRI deformation. *PloS One*, *12*(3), e0173372.
23. Zhang, Y. T., & Liu, S. Q. (2018). Individual identification using multi-metric of DTI in Alzheimer's disease and mild cognitive impairment. *Chinese Physics B*, *27*(8), 088702.
24. Sheng, J., Wang, B., Zhang, Q., Liu, Q., Ma, Y., Liu, W., ... & Chen, B. (2019). A novel joint HCPMMP method for automatically classifying Alzheimer's and different stage MCI patients. *Behavioural Brain Research*, *365*, 210–221.

25. Gosztolya, G., Vincze, V., Tóth, L., Pákáski, M., Kálmán, J., & Hoffmann, I. (2019). Identifying mild cognitive impairment and mild Alzheimer's disease based on spontaneous speech using ASR and linguistic features. *Computer Speech & Language*, *53*, 181–197.
26. Bhandare, A., Bhide, M., Gokhale, P., & Chandavarkar, R. (2016). Applications of convolutional neural networks. *International Journal of Computer Science and Information Technologies*, *7*(5), 2206–2215.
27. Huang, Y., Xu, J., Zhou, Y., Tong, T., Zhuang, X., & Alzheimer's Disease Neuroimaging Initiative (2019). Diagnosis of Alzheimer's disease via multi-modality 3D convolutional neural network. *Frontiers in neuroscience*, *13*, 509.
28. Hubel, D. H., & Wiesel, T. N. (1968). Receptive fields and functional architecture of monkey striate cortex. *The Journal of physiology*, *195*(1), 215–243.
29. Fukushima, K. (1980). A self-organizing neural network model for a mechanism of pattern recognition unaffected by shift in position. *Biological Cybernetics*, *36*, 193–202.
30. Gu, J., Wang, Z., Kuen, J., Ma, L., Shahroudy, A., Shuai, B., & Chen, T. (2018). Recent advances in convolutional neural networks. *Pattern Recognition*, *77*, 354–377.
31. Payan, A., & Montana, G. (2015). Predicting Alzheimer's disease: A neuroimaging study with 3D convolutional neural networks. *arXiv preprint arXiv:1502*.02506.
32. Aderghal, K., Boissenin, M., Benois-Pineau, J., Catheline, G., & Afdel, K. (2017, January). Classification of sMRI for AD diagnosis with convolutional neuronal networks: A pilot 2-D+ ε study on ADNI. In *International Conference on Multimedia Modeling* (pp. 690–701). Springer, Cham.
33. Vu, T. D., & Yang, H. (2017). Detecting Alzheimer's disease by sparse autoencoder and convolutional network on multimodal data pp. 278–281.
34. Luo, S., Li, X., & Li, J. (2017). Automatic Alzheimer's disease recognition from MRI data using deep learning method. *Journal of Applied Mathematics and Physics*, *5*(9), 1892–1898.
35. Khvostikov, A., Aderghal, K., Benois-Pineau, J., Krylov, A., & Catheline, G. (2018). 3D CNN-based classification using sMRI and MD-DTI images for Alzheimer disease studies. *arXiv preprint arXiv:1801.05968*.
36. Tang, H., Yao, E., Tan, G., & Guo, X. (2018, August). A fast and accurate 3D fine-tuning convolutional neural network for Alzheimer's disease diagnosis. In *International CCF Conference on Artificial Intelligence* (pp. 115–126). Springer, Singapore.
37. Feng, C., Elazab, A., Yang, P., Wang, T., Zhou, F., Hu, H., ... & Lei, B. (2019). Deep learning framework for Alzheimer's disease diagnosis via 3D-CNN and FSBi-LSTM. *IEEE Access*, *7*, 63605–63618.
38. Fu'adah, Y. N., Wijayanto, I., Pratiwi, N. K. C., Taliningsih, F. F., Rizal, S., & Pramudito, M. A. (2021, March). Automated classification of Alzheimer's disease based on MRI image processing using convolutional neural network (CNN) with AlexNet architecture. In *Journal of Physics: Conference Series* (Vol. 1844, No. 1, p. 012020). IOP Publishing.
39. Al-Khuzaie, F. E., Bayat, O., & Duru, A. D. (2021). Diagnosis of Alzheimer disease using 2D MRI slices by convolutional neural network. *Applied Bionics and Biomechanics*, *2021*, 6690539.
40. Aderghal, K., Benois-Pineau, J., Afdel, K., & Gwenaëlle, C. (2017, June). FuseMe: Classification of sMRI images by fusion of deep CNNs in 2D+ ε projections. In *Proceedings of the 15th International Workshop on Content-Based Multimedia Indexing* (pp. 1–7).

10 Deep Learning Applications on Edge Computing

Naresh Kumar Trivedi, Abhineet Anand, Umesh Kumar Lilhore, and Kalpna Guleria
Chitkara University Institute of Engineering & Technology, Chitkara University
Punjab, India

CONTENTS

DOI: 10.1201/9781003143468-10

10.1 INTRODUCTION

Innovations and the widespread adoption of high-speed mobile networking in personal computing devices mean that today's crowd-sourced applications are distributed globally, and crowd-sourced data is rather heterogeneous. Crowd-sourced applications obtain resources by collecting multiple contributors' raw data. They must also be deployed on cloud platforms for self-service on-demand, unlimited pooling of resources, and elastic scalability. In various crowd-sourced applications, including spoken recognition, recommendation systems, and video classification, modern deep learning techniques have rapidly gained popularity. Nevertheless, there is significant pressure on the infrastructure in the most modern cloud computing paradigms given the high amount of crowd-sourced data and high requirements in the conventional deep learning processes, such as facial recognition or monitoring human resources in camera networks. Edge computing recently suggested that cloud computing be supplemented by other data processing operations at the network's edge. This last paradigm generation demonstrates a significant reduction in machine time, memory costs, and energy use compared to traditional cloud computing over a wide array of Big Data applications [1].

Deep learning is prevalent in various areas of use, such as computer vision, natural language processing, and Big Data analysis. For example, in many computer vision competitions over the past few years, deep learning approaches have consistently outperformed conventional object recognition and detection methods. However, the high degree of precision in deep learning comes at the cost of increased computation requirements and storage for both training and learning phases. Training a profound learning model is costly in space and computer resources because it iteratively refines millions of parameters over many periods. The results are computationally expensive because the input data (e.g. a high-resolution picture) is theoretically significant, and the input data requires millions of calculations. The characteristics of deep learning are that it is highly precise and requires a high use of resources [2].

1. **Latency:** For several applications, real-time deduction is essential. Autonomous camera frames, for example, need to be rapidly processed to identify and prevent obstacles. A voice-based assistance application must search, understand, and respond quickly to user requests. However, sending cloud data or training data may result in more network queuing and propagation delays and it does not meet strict bottom-to-bottom low-speed requirements for immersive real-time applications. Actual studies have shown, for example, that more than 200 camera frames need to be downloaded to Amazon Web Services and view work carried out [3].
2. **Scalability:** The transition of data from the sources to the cloud causes scalability problems when the number of connected devices grows when a network connection to the cloud becomes a bottleneck. Network resources are also ineffective for downloading all data to the cloud, particularly when deep learning does not need all data from all sources. Bandwidth-intensive data sources, such as video streams [4], are critical.
3. **Privacy:** Sending data to the cloud poses privacy problems for users who own or store data. Users can carefully upload their cloud information (e.g., face or language) and how the cloud or application uses them. The recent use of cameras and other sensors in a smart city, such as in New Mumbai City, for example, posed a significant concern [5–7].

Edge learning, a complementary service for existing computer platforms, is dedicated to addressing the challenge of crowd-sourced, deep learning applications to address the huge network traffic and high computational requirements as well as increase device response times. Edge learning performs data preliminary network learning; raw information in local regions is processed on edge servers to minimize network traffic so that computation in data centers is accelerated.

By 2024, it is estimated that 5G mobile edge computing (MEC) will be a multimillion dollar market with $73 million in company deployments. The sophistication of the data keeps growing every year. The increase in network complexity systems results from the increase in on-demand and personalized services. Providers of internet services have to handle traffic in connected cars, online gaming, voice over IP (internet protocol), and IoT computer transmissions on the internet for web browsing. A fundamental change in mobile and fixed access networks is necessary for new restrictions that are being imposed by on-demand providers, as listed previously. Mobile networks of the fifth generation (5G) are being built to meet the growing demand and diversity [8].

Network providers use cloud computing techniques to cope with the dynamic traffic requested by modern users. In order to minimize the operating costs of increasing mobile networks to provide on-demand service, 5G networks are using Software-Defined Networking (SDN) and virtualizing network functions. In the long term, end users should anticipate improvements to be made to work because 5G is optimized to offer low latency, delays, high capacity, and high bandwidth communications for various applications, such as autonomous vehicles and robotics of Industry 4.0.

Since 5G networks have tight technical specifications, designers have rethought the backbone and access network architectures to better accommodate core functions and complex networks. The advent of mobile edge computing disrupts the conventional distinction between end user network connectivity and core network reliability (information computing and storage).

The combination of mobile edge computing and cloud computing gives rise to virtualized network extensions and management planes that extend it to end-to-end services. Mobile edge computing will then serve as an essential component of the network transformation, such as quality of service and power usage. Mobile edge computing, in conjunction with SDN, was described by the European 5GPP (a Private-Public Partnership) as a major enabler of meeting throughput, scalability, latency, and automation requirements in 5G [9, 10].

Mobile computing puts computational power that is accessible to the consumer closer to the transaction. This proximity reduces the amount of traffic across the core network, increases response speed with latencies of ten milliseconds, and helps make certain computational tasks, such as online machine learning (ML), easier to handle in the cloud. The provision of cloud-based computing and network operators will allow a real-time investigation through location-based collaboration [11].

To infrastructure management, device, service, and network administration, the demands of heterogeneous connections pose a dynamic challenge in 5G networks. One of the main advantages of mobile edge computing is delivering, orchestrating, and managing services over time and space, which shift in place and location. A promising approach uses ML to satisfy this new set of demands, which conventional optimization methods can't cope up. The core network and ML techniques for processing in mobile networks need to evolve. In real-time networks, current network optimization techniques are too slow to keep up with the required research. Over the past two decades, ML has gained fame for pattern recognition [3, 12].

Formal artificial intelligence (AI) methods have been studied and used in computer vision and natural language processing shown in Figure 10.1. Cognitive automated 5G automatically configures, optimizes, secures, and recovers while augmenting real-time response complexities; however, it comes with extra overhead, capital expenditure, and resource usage. Cloud-based technology and mobile edge computing working together would improve resource use and profits and make service providers more energy efficient [12, 13].

Deep learning is a type of ML and is an artificial intelligence subset on the other side. AI is a general term referring to techniques that enable computers to imitate human conduct. ML is a series of data-trained algorithms that allow all this.

But deep learning is nothing more than ML based on the human brain's structure. In the context of a given data structure, a deep learning algorithm tries to draw the

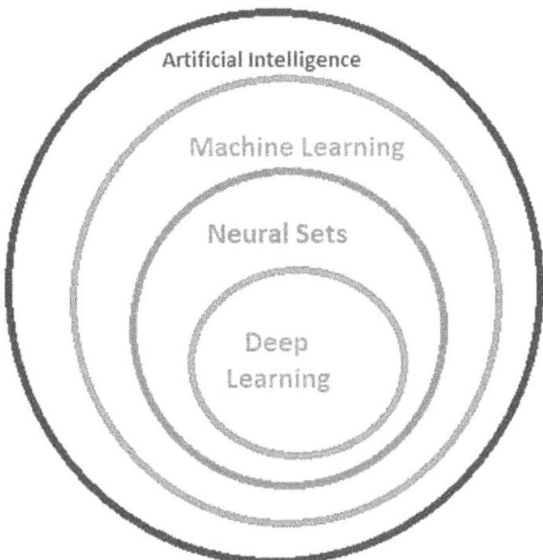

FIGURE 10.1 Deep learning.

same conclusions. Deep learning uses many algorithms, called neural networks, to achieve this aim. Based on the brain anatomy configuration, neural networks were developed. Neural networks can be trained to recognize patterns and to differentiate between a different kind of knowledge. They can see a neural network as a coarse to the fine layer, which increases the chance of accurate result detecting and outputting. The architecture of the human brain works similarly in our minds. Whenever we have new information, our brains attempt to match it to known objects to check it. In other words, deep neural networks enable us to perform multiple operations, such as grouping, trial, and regression [14, 15].

Neural networks may carry out the same tasks as classic ML algorithms. The opposite is not true, though.

Scalable deep learning services are the basis for many limitations. Depending on your target application, you can expect low latency, improved security, or long-term economic performance. Under such conditions, maybe the best way to host a deep learning model is to use deep-edge learning models that provides additional benefits. Edge applies here to the local consumer goods estimate. Edge applies here to the local consumer goods estimate.

10.1.1 Why Edge Is Preferred

The reasons for favoring edge computing over cloud computing are abundant. A few of its properties are shown in Figure 10.2.

10.1.2 Latency and Bandwidth

Indeed, a remote server that runs a time trip is called by the API (RT). Applications requiring an almost immediate deduction cannot function properly when they are sluggish. For example, if the latency is high enough in auto-driving vehicles, the risk

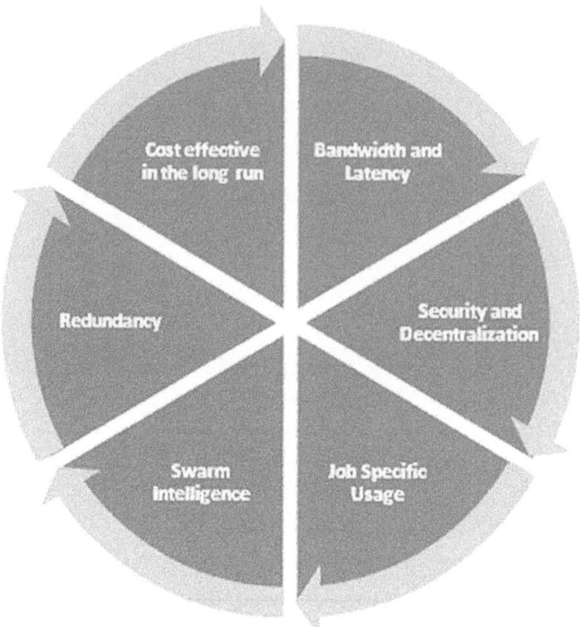

FIGURE 10.2 Properties of edge computing.

of injuries can increase substantially. Also, specific animation frames will feature rare incidents such as animal crossings and jaywalkers. Reaction times are essential to satisfy the needs of consumers like these. That's why customized Nvidia hardware is used for edge inferences [12].

On the contrary, adequate bandwidth is significantly reduced if you have multiple devices on the same network. Communication channel competition exists when it is done on the outside; the computation can be substantially decreased. Many optical videomaker recorders are in 4K, which saves a lot of bandwidth, like smartphones, laptops, camcorders, and TVs. That's because we don't have to give the cloud data. We can efficiently expand this network by not sharing the cloud.

10.1.3 Decentralization and Security

Business servers can be hacked and attacked. Of course, if you use a trustworthy seller, the danger is negligible. However, you must trust a third party to ensure that your data and IP are protected. By getting computers on the side, you get complete control of your IP.

10.1.4 Job-Specific Usage

You're going to have a toy factory. It has hundreds of jobs. An image classification service is required at each workstation. In each workstation, various artifacts may not work out a single classifier. Moreover, hosting a variety of cloud classifiers will be very expensive. The system's economy is that individual cloud classifiers are trained to be sent to the edges of the trained models. These computers have now been adapted for your desktop. They will be better than a classification that forecasts every workstation.

10.1.5 Swarm Intelligence

Further to the above concept, edge devices can also support computer models in training. This is particularly useful for strengthening education, for which a large number of "episodes" can be simulated in parallel. Besides, online learning data can also be collected using edge computers (or continuous learning). For example, multiple drones can be used to survey a classification area. A single model with optimizing techniques such as the asynchronous Stochastic Gradient Descent (SGD) can be trained simultaneously between all edge equipment. It can also be used only for aggregating and processing data from various sources.

10.1.6 Redundancy

For network architectures, robust memory and redundancies are highly vital. If one node in a network fails, the other nodes may have significant impact. Edge systems will provide a decent degree of redundancy in our case. If one of our edge devices fails (a node here), your neighbor takes over temporarily. This guarantees reliability and significantly decreases downtime.

10.1.7 Cost-Effective in the Long Run

Eventually, cloud services' cost is higher than a certain number of inferior computers. This applies especially to your devices' large duty cycles (i.e., most of the time they work). Also, edge equipment, when produced in bulk, is considerably cheaper, minimizing costs.

10.1.8 Limitations for Deep Learning on the Edge

Profound models of learning are famous for their size and expense. This is a challenge that is usually frugal to fit into edge devices. These problems can be dealt with in various ways as shown in Figure 10.3.

10.1.8.1 Parameter-Efficient Neural Networks

Their massive scale is a striking feature of neural networks. Usually, edge devices cannot accommodate large neural networks. This led researchers to reduce the size of the neural networks while preserving accuracy. MobileNet and SqueezeNet are two common neural efficiency parameters [16].

The SqueezeNet contains many techniques, including late sampling and filter reduction, to achieve high efficiency at a low level of parameters. They add "Fire modules," which have network parameter efficiency "squirting" and "expanding" layers.

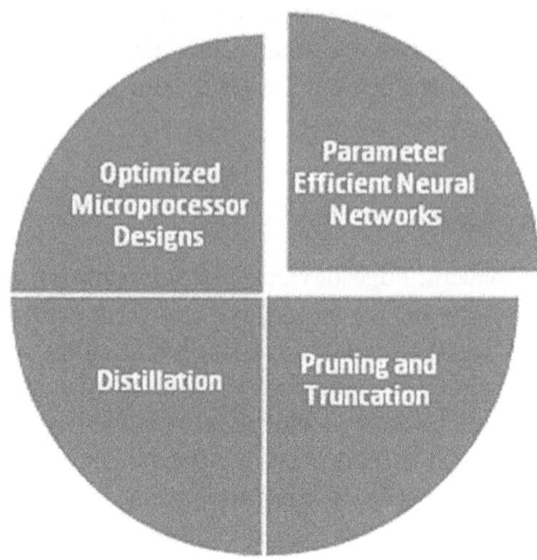

FIGURE 10.3 Limitation for deep learning on the edge.

10.1.8.2 Pruning and Truncation

Many neurons are benevolent and do not contribute to the absolute exactness of the trained network. In this case, we can save space by plucking these neurons. Learn2Compress by Google has shown that we can reduce the factor by 2 while preserving the accuracy of 97%.

Also, 32-bit float values are the most neural network parameters. On the contrary, edge devices may be configured to operate at or below 8-bit values. The model size can be considerably reduced by reducing accuracy. The reduction of the 32-bit model to the 8-bit model preferably decreases the model's size by one fourth.

10.1.8.3 Distillation

Distillation teaches smaller networks into a giant network of teachers. This is used in the sizing of Google's Learn2Compress. When combined with transfer learning, this is an extremely effective method for shrinking the model's size without sacrificing accuracy.

10.1.8.4 Optimized Microprocessor Designs

We also addressed how neural networks can be stretched to accommodate our edge devices. The performance of the microprocessor will be measured as an alternative (or complementary) method.

A microprocessor, like the popular Nvidia Jetson, would be the easiest solution. However, when implemented on a large scale, these devices cannot be economically successful.

10.1.9 Application of Deep Learning Using Edge

A few years ago, we never expected profound apps to carry self-driving cars and virtual helpers like Alexa, Siri, and Google Assistant. Now, though, these inventions are part of our daily lives. Deep learning keeps us fascinated by its infinite options for detecting fraud and pixel restoration. Let us understand deep learning applications in all industries.

Top applications of deep learning across industries are given in the next section.

10.1.9.1 Fraud Detection

The holy grail of businesses has long been identifying and stopping fraud in real time. For some time, technology has provided extremely robust tools to prevent fraud. Advanced analytics and artificial intelligence move things to a new stage—well before fraud occurs.

According to an interview of 1,055 managers by the Association of Certified Fraud Examiners and SAS, advanced analytics and biometrics are becoming the main focus of anti-fraud programs. The study found that 13% of companies use AI and machinery to detect and prevent fraud. Another 25% expect to implement AI and machinery for the next year or two—an increase of 300% [5].

Biometrics—fingerprint, facial, or keystroke recognition—are the most popular methods for preventing fraud, with more than a quarter using such techniques. At the same time, another 16% expect biometrics to be used in the next two years.

Blockchain/distributed ledger technology and robotics are less common than biometrics, including robotics process automation (9% of organizations for both categories). However, in the next two years, large numbers of companies expect these innovations to be implemented.

The least used technology is virtual or augmented reality in anti-fraud programs. Today, just 6% of organizations use this technology, and almost two-thirds do not plan to.

- Nearly 35% (72%) of organizations intend to use automatic tracking, exception reporting, and the identification of anomalies by 2021. Likewise, nearly a half of all companies plan to use predictive analytics/modeling (52% up from 30 percent), and data shows (47%; currently 35%).

10.1.10 Deliver a Quality Experience at Scale

Limelight has the option to install and execute network edge video applications. You may pick locations worldwide to fit your geographical access and scale requirements. Various cutting calculation options allow you to simplify components that add delays, expenses, and inefficiencies to your service architecture.

Moreover, you can boost your delivery by incorporating edge computing in your content delivery network (CDN) workflow using CDN services from limelight. Using cached locally is much more effective than using a centralized role by ingesting multi-edge Point of Presence (POP) and processing locally. Using one vendor also simplifies sourcing, workflow, and operations.

The latency-sensitive workflow, including videos, is suitable for limelight edge calculations. Implementing edge will reduce the average latency dramatically over centralized computation approaches. The management of workloads at the edge will increase latency, efficiency, and reliability in use cases involving a CDN.

10.1.11 Cost Management

In addition to improving latency and size, edge calculations implementation will allow you to reduce costs associated with centralized cloud services, storage, and network transport. In the CDN case, you can achieve additional cost optimization with just one provider. Limelight offers zero-rate traffic from edge computes to our CDNs.

10.2 LIMELIGHT EDGE COMPUTE OFFERINGS

Limelight provides the best possible solution for your particular environment with its collection of cutting-edge calculation solutions, which are discussed in the next sections.

10.2.1 Bare Metal

Limelight Bare Metal Service (BMaaS) gives you calculating power, where it is needed without resource sharing. Bare metal is ideal if complete control and optimum efficiency of computational resources are needed.

10.2.2 Virtual Machine

As a solution, Limelight Virtual Machine (VM) provides a virtual computer capability that is easy to deploy worldwide and easy to measure and develop as computing requirement change.

10.2.2.1 Serverless Compute

Limelight edge functions, integrated tightly with the CDN, automates the code for many leading sites worldwide and executes next to your users at your network edge to ensure the lowest latency and demand size.

Limelight's advanced calculation provides benefits from direct peer connectivity for more than 1,000 ISPs and cloud providers, thus eliminating much latency and confusion in relying on conventional public internet routing.

10.2.2.2 Healthcare

10.2.2.2.1 Rural Medicine

Historically, it has been a struggle to ensure good healthcare in remote rural areas. Even today, medical suppliers struggle to provide quick, quality healthcare to people living far from hospitals and with limited access to the internet with innovation in telemedicine and more easily accessible health data. Traditional healthcare databases face significant challenges because of connectivity problems, but the combination of IoT medical devices and cutting-edge computer applications will simplify these problems.

Portable IoT healthcare equipment built by cutting-edge computer companies is in the position, without continuous interaction with network infrastructure, to capture, store, and analyze sensitive patient information. Patients with wearable IoT devices can be quickly and efficiently diagnosed on-site; and when connections are restored, the information collected can be returned to central servers. IoT healthcare systems can expand the scope of existing networks by connecting with an edge data center to allow physicians to access critical data, even in weak connectivity areas. This is only one of the edge cases in which the scope of healthcare facilities can be extended [17].

10.2.2.2.2 Patient-Generated Health Data

Patient-Generated Health Data (PGHD) is a rapidly growing field where the availability and development of the technologies have, in many instances, outpaced the publication of trials designed to evaluate health outcomes, usability, interoperability, and benefits and harms of these technologies. Therefore, it is essential to determine which available technologies have been assessed to determine efficacy related to health outcomes for consumers with (or at risk for) chronic diseases.

P. Subramaniam and M. J. Kaur report focuses on consumer technologies that provide PGHD. These are commercially available devices to consumers and do not require a prescription from a physician. Therefore, this report does not include medical devices that perform remote patient monitoring, which falls more broadly within telehealth [18].

10.2.2.2.3 Improved Patient Experience

Going to the doctor should not be difficult or irritating. IOT medical devices are among the most critical computing devices that improve the patient experience fully in the healthcare industry from smart devices that allow people to check appointments whenever they wish to receiving alerts that direct them through a new facility to find the correct office.

Advanced computers can take on a much more integral role in IT infrastructures in the healthcare industry, with many IoT medical devices supporting patients and improving the experience. Many hospitals offer patients streaming entertainment facilities from movies and sports to immersive educational programs. Edge data centers can help to decentralize content with reduced latency and provide it more broadly [19].

10.2.2.2.4 Supply Chain

The supply chains in manufacturing are one of the most convincing examples of edge computing in action. Today's hospitals and clinics are technological marvels, outfitted with cutting-edge medical technology and computer hardware in order to provide the best possible care. They also have less advanced medical equipment, that is no less essential, used to save lives in daily procedures. It is a huge logistical challenge to keep these facilities operating. Any interference with the supply chain that keeps them working creates significant health risks from costly mechanical elements for robotically assisted surgical tools to the smallest bandage.

IoT edge devices with sensors will revolutionize the management of medical facilities' inventories. Devices that collect data on the patterns of use can use predictive analysis to assess if hardware fails. At the same time, smart Radio Frequency Identification (RFID) tags can remove long-term paperwork and manual ordering from inventory administration. The location of vital shipments in real time can be monitored by floating vessels fitted with GPS and other sensors. IoT healthcare supply chain technologies provide an opportunity to achieve operational efficiency on the margins for organizations that struggle to manage rising costs and constitute one of the most significant cases for computer technology use [20].

10.2.2.2.5 Cost Savings

Regarding cost savings, analysts expect that the widespread use of IoT cutting devices will enable healthcare organizations to reduce their business costs by up to 25%. Some of these savings would come from daily applications, such as safety and surveillance or intelligent building inspections, but patients monitoring and interaction could provide real innovation. Wearable IoT medical devices and implantable sensors are among the most advanced computer applications that could significantly reduce patient costs through the entire treatment cycle.

Another possible source of cost savings is interconnectivity. Medical service providers have been plagued for a long time by inconsistent structures and difficult recording, all but eliminated by IoT medical device networks and cutting-edge computing applications that can interact rapidly and efficiently between organizations. Any technology that can increase productivity and provide a better product is undoubtedly rapidly implemented, with growing costs posing a continuing challenge to people's access to health services.

While IoT edge devices are already trendy, their full potential has not yet scratched the surface. As the quantity of devices is still growing and network database infrastructures are additionally burdened, advanced computation examples will soon be found in all medical IT strategies. The health sector will enormously benefit from both advances, and the one-two punch of IoT medical devices and edge computing will certainly bring key benefits in the future [21].

10.2.2.2.6 Self-Driving Cars

- Milliseconds matter when you drive a car. Autonomous vehicles are no different, even though your AI powers them. AI = data + calculator, and you want your estimate as similar as possible to your data.

10.2.2.2.7 Computing in Edge

- People are both mindful of the cloud and love it. How about not having to worry about what your gadgets will do and about having almost unlimited elastic storage and calculation power?
- Okay, a couple of things. In the end, the cloud is just another's machine, as the aphorism goes. Okay, there are maybe millions of computers in superpowerful data centers, organized with respect—but these are all computers belonging to someone else.

- However, does it matter if someone can provide all that you need, potentially more effective than your own company, along with security guarantees? It doesn't in certain situations. But when it comes to autonomous vehicles, it is essential.

10.2.3 Autonomy and Cloud Don't Go Well Together

Let's look at the notion of autonomy to understand why. Autonomy is characterized as "independence or freedom by the will or the action." If you rely on someone else's machine, can you be autonomous? Not so. Not so.

Yes, redundancy exists, and yes, even service-level agreement (SLA) can exist. But when all is said and done with the cloud, you usually connect to another device over the internet. What happens if you run into connection problems while you are in a moving vehicle and this vehicle depends on a cloud-based computer for its main functions?

This is not the same as loading your favorite cat photos and they are lagging. A lag is a matter of life and death in a moving vehicle situation. And in cases like this, what can be done? Computing in edge.

Edge computing is a phenomenon where data is produced outside the data center, and the data is measured as similar to the data as possible. This means fragile, prefabricated data centers in real life.

Naturally, small is a subjective concept. Is a container small? Perhaps if you equate it with a data center, such as a cloud provider. However, in our houses, it's not something that any of us could or will have.

However, we still have homes where some of the significant edge computing capabilities are used. For smart homes or smart city situations, connected devices that communicate through IoT sensors are an excellent match for cutting-edge computing. These scenarios, wholly blown, involve many devices gathering and sharing a large amount of data.

10.2.3.1 Language Translations

In November 2016, Microsoft benefited developers and end users with its AI-based machine translation known as Neural Machine Translation (NMT). By leveraging neural processing unit (NPU), an AI-dedicated processor built into Mate 10, Huawei's new flagship handset, Microsoft took NMT functionality to the edge of the cloud. Even without internet access, this new chip provides AI-powered translations for the device, which enable the machine to produce translations that are in line with the online system's quality. Microsoft and Huawei researchers and engineers have worked together to adapt the neural translation to this new computer world to achieve this breakthrough [22].

The most sophisticated NMT systems in development at present (i.e., those used by companies and applications to scale in the cloud) use a neural network architecture that combines many layers of Long Short-Term Memory (LSTM) networks, a care algorithm, and a decoder layer of translation.

10.2.4 AutoML Translation

Developers, translators, and localization professionals can easily create production-ready, high-quality models with minimal ML expertise. Translated language pairs are uploaded, and AutoML Translation builds a custom model adapted to particular domain requirements.

10.2.5 Translation API

Translation API Basic translates a text for your website or mobile app between more than 100 dialects in real time. Advanced Translation API has the same quick and interactive results as Basic Translation API and additional customization options. For domain and context-specific terms or phrases, customization is important.

10.2.6 Media Translation API

With increased precision and easy integration, the Medium Translation API offers real-time audio translation directly to your content and applications. You can also boost your user experience through low latency streaming translation and quickly internationalize your business.

Amazon Translate is a neural network machine translation service that delivers a high quality, accurate translation at an affordable cost. Neural machine translation is a type of automation in language translation that uses profound learning models to translate more precisely and sound more natural than statistical or rule-based algorithms in translation.

Amazon Translate enables you to quickly find content for your diverse customers, including websites and applications, translating vast amounts of text for analysis and allowing cross-language communication among users [13].

Recently, Amazon declared Translate the top machine translation provider in 2020 across 14 language pairs, 16 industry sectors, and 8 content types.

10.2.7 News Aggregation and Fraud News Detection

10.2.7.1 A Short History of Content Aggregators

The year 2015 saw the emergence of content aggregators in the digital industry. Since multiple publishers saw syndication as a sin, they opposed aggregators and chose Google News, Yahoo News, and AOL. Neither of them wanted to dig into their budget. Therefore, using Huffington Post, BuzzFeed, Vine et al., we're searching for free news instead.

The size of the audience at these aggregator sites was small, but it was believed that they could expand quickly. Many editors tried to achieve the common aim of "being your readers all over," and sadly, instead of seeking a better way for the public to read comprehensive news, began to offer their contents free of charge to those aggregators.

In return, they have plenty of content views, but not a great deal of return.

When you are concerned about sales, which is a considerable obstacle for digital publishers, you can only bet on subscription models and nothing else for content aggregators. This does not propose a win-win scenario for digital publishers [21].

10.2.7.2 The Impact of Content Aggregators on News Consumption

In the last decade, content discovery has changed considerably, but news consumption has changed dramatically too. Consumers demand high-quality sharable, searchable, and readily accessible news content that can be delivered frictionless in exchange for their interest.

Many users now opt to pay once a month for an aggregator platform to access news from several publishers. Paying once and consuming complete material during the month is more convenient and cost-efficient.

10.2.8 Fake News Detection Using Edge Computing

Over the years, fake news has spread rampantly on social media. Fake news has become a famous demon that affects the population overall. Its daily users are not only concerned, but also the advertisers are concerned about the effects of fake news on business. Online news outlets are a shield with two rims. Fake news is becoming a threat to our culture more and more. It is commonly used to draw audiences and also to increase advertisement profits for business interests. However, it has been known that media giants with potentially sinister motives generate false news to manipulate global affairs and policies.

Fake news from sarcastic posts to fabricated information and government misinformation in certain news outlets today creates various problems. There are widespread issues with the significant social implications of fake news and a lack of confidence in the media. Of course, "fake news" is intentionally deceptive, but the social media debate has lately changed its meaning. Some now use the word to reject evidence that contradicts their favorite points of view.

In particular, after the U.S. presidential election, the role of misinformation was a significant concern in American political discourse. The word "fake news" was used particularly to describe factually inaccurate and misleading stories papers often published to make money from page views. The law aims to generate a model that precisely forecasts the probability of false news in a given post.

After the media coverage, Facebook was at the center of much criticism. Model introduced a function to report false news on the website when a user sees it and has publicly said that they work to automate the distinction between these posts. This isn't a simple job, of course. A specific algorithm must be respectfully neutral, as false news on both ends of the spectrum exists and must compare credible news outlets on both ends of the spectrum. Furthermore, it is a complex question of validity. However, it is required to get an understanding of what fake news is to address this issue. Later, it is necessary to examine how ML and natural language processing help us identify fake news [23].

10.2.9 Natural Language Processing

To grasp and control the human language, natural language processing (NLP) uses algorithms. This technology is one of ML's most commonly used fields. As

AI expands, so does the need for building model professionals who analyze language and voice, unlock contextual trends, and generate insights into text and audio.

The encoding of NLP is one of the most critical information age technologies. A key element of artificial intelligence is understanding complex language statements. NLP applications are everywhere, and most people communicate in language: web search, ads, e-mail, customer care, translation of languages, radiology reports, etc. There is a wide range of essential tasks and learning machines for NLP applications. Recently, deep learning methods in several different NLP tasks have achieved very high efficiency. These models can also be trained using a single end-to-end model and require no standard functionality. Students will learn to implement, practice, debug, imagine, and invent their model neural network this spring.

We may therefore infer that NLP is the computer science subfield that focuses on computer comprehension and processing human language. The critical task of NLP is technically to program computers that analyze and process large quantities of natural language data.

10.2.10 History of NLP

It is possible to split the NLP history into four phases. The phases have different types of problems.

10.2.10.1 First Phase—The Late 1940s Until Late 1960s

The main subject of the work at this point was machine translation (MT). It was a time of excitement and hope.

The NLP research started in the early 1950s after a Warren Weaver's essay on the 1949 Machinery Translation Memorandum (MTM) study by Booth & Richens. In 1954 the Georgetown-IBM experiment demonstrated a minimum automatic translation experiment between Russian and English. *The MT Journal* was published in the same year. The journal for machine translation in 1952, the first Machine Translation (MT) International Conference was held, and in 1956, a second conference was held. In 1961, the work on Machine Language Translation and Analysis at the International Teddington Conference took center stage, and this method was then presented.

10.2.10.2 Second Phase—The Late 1960s to Late 1970s

The work done in this process was mainly linked to world awareness and its role in the creation and manipulation of representations of significance. This process is often referred to as the AI-flavored phase. The tasks in managing and developing data or information databases started at the beginning of 1961—AI inspired this work. Also created a question-response system for baseball in the same year. The system entry was limited, and the language processing was quick.

Marvin Minsky identified a highly sophisticated device in 1968. This method is understood and designed to determine the knowledge base in the interpretation and response of the linguistic information compared to the BASEBALL question answering system.

10.2.10.3 Third Phase—The Late 1970s to Late 1980s

The grammatical logical process can be defined as this phase. In the last stages, the researchers could not implement logic to reflect information and reasoning in AI because of the failure of the practically constructed method.

By the end of the decade, the grammar-logical approach helped with its strong general-purpose phrase processors, such as the SRI core language engine theory and representation of discourses, which offered a means of approaching broader speech. In this period, we gained some valuable tools, such as parsers; e.g., Alvey Natural Language Tools; and more business processing queries; e.g., for database queries. The 1980s also directed lexicon work toward grammatical logic.

10.2.10.4 Fourth Phase (Lexical & Corpus Phase)—The 1990s

It's a lexical process; we can understand it. In the late 1980s, the stage had a lexicalized approach to grammar. In this decade, the emergence of machine language learning algorithms led to a revolution in managing natural languages.

10.2.11 Virtual Assistants

A separate contractor is a virtual helper that provides administrative support to customers while operating outside the client's office. A virtual assistant usually works from a home office but can access the necessary planning documents from afar, such as shared calendars.

Individuals who work as virtual assistants also have several years of management experience. New opportunities will be provided to digital assistants with social media experience, content management, blog post writing, web design, and internet marketing. The demand for skilled virtual workers is expected to increase because employees and employers are more involved in homework.

A virtual assistant is an individual employee who specializes in delivering customer administrative services, typically at home from a remote location.

A virtual assistant can perform traditional tasks, including scheduling appointments, making telephone calls, handling travel agreements and managing e-mail accounts.

Some automated helpers provide graphics, blogs, bookkeeping, social networks, and marketing services.

The ability for an employer to contract only for the services they require is a significant benefit of employing a virtual assistant [24].

10.2.11.1 How a Virtual Assistant Works

Digital helpers have become increasingly popular, with small companies and startups relying on virtual offices for cost management. As an independent contractor, a virtual assistant will not receive the same pay and benefits as an employee working in an organization full time.

As the virtual assistant is working off-site, the organization's office does not need any desk or any other workspace. A virtual assistant must pay for and provide their own computers, standard applications, and high-speed internet services.

10.2.12 Visual Recognition

Edge computing contains computing (and some data storage) to devices that generate or retrieve data on a cloud-based central infrastructure (particularly in real times). There are no latency issues with this data, and this approach reduces transmission and treatment costs. It's as if it is all done on a laptop instead of in the cloud locally.

Since IoT devices with an internet connection have increased exponentially, advanced computing is generated for either cloud information. And many IoT devices produce a lot of information during operation.

Edge computing poses new possibilities in object detection, IoT (and face), language processing, and obstacle prevention, especially for those relying on ML. Image data is an excellent addition to IoT and a significant user of resources (power, memory, and processing). "Edge" picture processing, running AI/ML classics models, is a giant leap!

Express is one of the programs used by most researchers/researchers nowadays in the area of visual recognition [25, 26].

10.2.13 TensorFlow Lite

TensorFlow Lite is an open-source deep learning framework for small, small-latency computer ML. It is built to support you "at the edge" of your network instead of sending back and forth data from a server.

There are two main components in TensorFlow Lite (TF Lite):

The TF Lite transformer is transforming TensorFlow models into an efficient interpreter type and optimizing binary size and performance.

The TF Lite interpreter works on various hardware, including smartphone, embedded Linux computers, and microcontrollers, with specially optimized versions.

In short, the qualified and stored TensorFlow (like a model h5), which will be used by the TF Lite Interpreter inside an edge (like a Raspberry Pi), will translate into a TFLite FlatBuffer (like model Tf Lite).

For instance, in a Mac (the "Server"), we trained from scratch the simple CNN image grading model. The last model was trained with 225,610 parameters using CIFAR10 as input: 60,000 pictures (shape: 32, 32, 3). The model trained was 2.7 MB in size (iffar10 model.h5). With the TFLite converter, 905 KB (approximately 1/3 of the original size) was applied to the Raspberry Pi model (model cifar10.tflite). Deduce the results from both (.h5 on Mac and tflite on RPi), the same results are obtained [19].

10.2.14 Detecting Developmental Delay in Children

You child is experiencing a developmental delay if they fall under one or more areas of social, behavioral, or physical development behind their peers. Early care is the best way to help your child improve or even catch up if your kid is delayed. In babies and young children, there are several different kinds of developmental delays. They include issues with:

- Language or speech
- Vision

- Movement—motor skills
- Social and emotional skills
- Thinking—cognitive skills

There are periods when two or more of these areas are greatly delayed. It is known as "economic development delay" when this occurs. It applies to children and preschool children up to five years of age who have a minimum delay of six months. A developmental delay is distinct from disorders such as brain paralysis, hearing loss, and autism, which generally last for life.

Kids learn to crawl, chat, or use the toilet at various speeds. However, some children hit the milestones much later than other children. These delays have many causes, including:

- Premature birth
- Genetic disorders such as Down syndrome or muscular dystrophy
- Impaired vision or hearing
- Malnutrition
- Prenatal alcohol or substance use
- Physical violence or neglect
- Inadequate oxygen throughout childbirth

A system uses a validated screening tool to identify developmental delays with appropriate precision. However, only 48% of pediatricians used a standardized developmental screening method in 2011. Fifty-two percent of parents reported informally about the growth of their children, and 21% reported by filling a questionnaire, according to a study from the Centers for Disease Control and Prevention.

Numerous barriers exist when it comes to checking for developmental delays in routine clinical practice. According to a study, 82% of primary care physicians cited time constraints as the primary impediment. Other screening hurdles include:

- Overlapping health needs
- Lack of accessible referral suppliers
- Personnel requirements
- Lack of agreement about the appropriate screening methods and lack of trust in medical training
- The ability to treat behavioral and emotional problems in children effectively

There is also a high turnover of employees with subsequent training in tool management and a lack of compensation.

Developmental screening tests cannot be used for a developmental disorder diagnosis; hence, a method must be used that is most precise to avoid detection and overreferences. The literature did not identify an ideal initial screening method. A perfect test will cover all growth areas, apply equally to all ages, and have structural validity and fewer false negatives and false positives. The American Academy of

Pediatrics (AAP) recommends systematic assessment methods for fine and gross motor abilities, language and communication, problem-solving, adaptive behavior, and personal-social skills. Culturally sensitive screening strategies should be assessed in the mother tongue of the patient.

10.2.14.1 Colorization of Black and White Images

Most pictures in the nineteenth century were monochromatic because people painted them by hand, mostly simply because of aesthetics. Color images looked better on the mantle. For these purposes, contemporary artists continue to add color. The artist and teacher Tina Tryforos started hand-coloring images before going digital. Early in her career, she worked by hand with a minimal palette of colors, because she prefers simplicity. "I love the more nostalgic and painterly look instead of trying to make it look precise," she says. [36]

Photographer Kenton Waltz enjoys black-and-white images so he can decide which colors are used in a picture and which are left out. "The simplicity of color is the elegance of a frame," he says. "You value light and dark consistency." [36]

None of the colorization channels you know is the colorization information of the RGB color channel inside the three channels if any one channel did not exist, which destroys the colors in the picture. You know the colorization information of the image.

You want to emit three channels of the RGB components of this black-and-white image, which is the critical issue when using this black-and-white image as an input. Now imagine that you have a black-and-white image and have placed it in the black box from the right side, resulting in three RGB elements, one of which is the black box auto-encoder in response.

10.2.15 What Is Auto-Encoders?

Ian Goodfellow, the inventor of GANs, described Auto-encoders in his 2016 book *Deep Learning* as follows:

> An auto-encoder is a neural network trained to copy the input to the output. Internally, it has a secret layer h defining the input code. The web consists of two parts: the h=f(x) encoder function and the r=g reconstruction decoder (h). Figure 10.4 illustrates this architecture. It doesn't really work if an auto-encoder will learn simply to set g(f(x))=x anywhere. An auto-encoder is instead designed to keep you from perfectly copying. Usually, they are limited so that they can only copy information roughly similar to the training data and only copy input. Since the model is forced to give priority to copying whatever type of the input, useful properties of data are often taught. The definition of an encoder and decoder has been extended to stochastic mappings, p encoder (h|x) and p decoder (x|h), beyond deterministic functions.
>
> Historically, auto-encoders have been used to reduce the number of dimensions or to learn new functionalities. Additionally, recent advancements in generative modelling have elevated auto-encoders to the forefront of theoretical relationships between the auto-encoder and latent variable models. Auto-encoders can be thought of as a subset of feedforward networks since they are trained using the same technology, usually using minibatch gradient descent following back-propagation measured gradients.

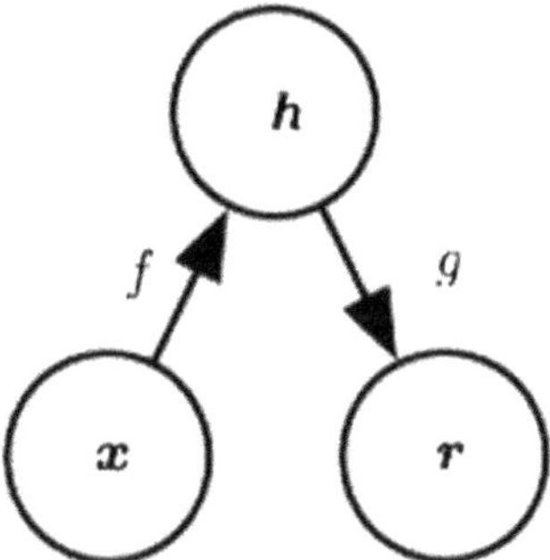

FIGURE 10.4 Auto encoder.

> The auto-general encoder's structure, as mapped by an internal representation or code h from an input x to an output (reconstruction). Both the encoder f(mapping x to h) and decoder g are composed of two components (mapping h to r). One solution is to create two copies, one of which will be a grey image. The input will be an encoder that extracts the functions of the photo "Latent Space Representations" image that can be used to reconstruct an image for the decoder. In contrast, the other copy will be the same image, but colored to represent the decoder's target.
>
> **Ian Goodfellow, Yoshua Bengio, Aaron Courville [27, 35]**

10.2.16 Adding Sounds to Silent Movies

Movies have often added sound effects not captured for more realistic purposes during the movie. This is a "Foley" operation. This method has been automated by researchers at the University of Texas. With 12 famous movie events, Jack Foley added Foley effects; he has developed a neural network [37]. Their neural network classifies the sound class to be created and also has a sequential sonic network. Thus, neural networks have used a whole different modality from temporally aligned images to sound generation!

The first thing the researchers did was construct a dataset containing short film clips with 12 film events (Automatic Foley Dataset). They created sounds inside a studio for some movie events (such as cutting, footsteps, and a clock sound). They downloaded video clips using YouTube sounds for other events (such as guns, horses running, and fire). The total length of 1,000 videos was five seconds.

The next move was to forecast the correct sound class. They compared two approaches for this purpose:

1. Frame sequence network (FSLSTM)
2. Frame relation network (TRN)

Every video frame was taken in the frame sequence network approach. They then interpolated frames for greater granularity between the current frames in the video.

The image features were derived from a ResNet-50 convolutionary neural network (CNN). Then the sound class was predicted with Fast-Slow LSTM-fed image features through a recurrent neural network. They attempted to capture detailed transformations and behavior of the objects in less calculative time in the frame relations network. The network structure relationship (the multi-scale temporal relationship network, more precisely) compared characteristics in frames at N-distances, in which N assumes many values. Finally, all these characteristics were combined with a multilayer perceptron again [28].

They took each video frame in the frame sequence network technique. They then interpolated frames between the video's current frames to add granularity. Image attributes were extracted using a ResNet-50 CNN. The sound class is then predicted using an image-based recurrent neural network called Fast-Slow LSTM. They attempted to capture the detailed transformations and behavior of the objects in the Frame Relation Network with the least amount of computational time possible. The frame relation network (or, more specifically, the multi-scale temporal relation network) compares features between frames separated by N, where N can take on any value. Finally, all of these characteristics were combined using a multilayer perceptron.

The results of the algorithm, including a human qualitative test, were analyzed with four different approaches. Local university students were asked to choose the sound that was the most natural, the most comfortable, the one with the least noise, and the most synchronized. In 73.71% of cases for one model and 65.96% of cases for another model, the synthesized sound was favored over the original sound. The choice depended on the video for each model: one model performed better on the scenes with several random changes in action [29].

10.2.17 Automatic Handwriting Recognition

By 2025, the Optical Character Recognition (OCR) market is expected to hit $13.38 billion; this rise of 13.7% is driven by the rapid digitalization of business processes using OCR to minimize labor costs and free up valuable work hours [30]. Whereas OCR has been viewed as a solution, a crucial factor, handwriting recognition or manuscript text recognition, continues to be viewed as a challenge. In comparison to typed text, the wide variety in handwriting styles between individuals and the poor quality of the writing pose major barriers to translating readable text to the machine. However, it is important to address for several industries, such as healthcare, insurance, and banking.

Deep learning techniques, such as transformer architectural growth, have made significant strides in recent years toward cracking handwritten text recognition. Smart character recognition is used to refer to handwritten text recognition because the algorithms required for ICR resolution require significantly more intelligence than generic OCR solutions [31].

10.2.18 Challenges in Handwriting Recognition

- High inconsistency and ambiguity between the strokes of a person to a person.

- The handwriting style of the individual often differs from time to time and is inconsistent.
- The text of the written documents is placed in a straight line, while the individual does not need to write a text line straight on a white paper.
- The cursive handwriting process makes separating and identifying characters challenging.
- Gathering a good labeling data collection is not inexpensive compared with synthetic data.

ML methods, including hidden Markov models (HMM), Support Vector Machine (SVM), etc., were the initial approaches to solving handwriting recognition. After pre-processing the initial text, the function extraction defines key details of an individual character, such as loops and inflection points, aspect ratios, etc. These features are then supplied to an HMM classifier to obtain the data. Due to the manual extraction process and limited learning ability, ML model's output is very restricted. The extraction step for each language varies and is therefore not scalable. With this deep learning, the accuracy of handwriting recognition has improved tremendously [32].

RNN(RN)/LSTM (long short-term memory networks) can handle sequential data in which time patterns are identified and produced. However, they deal with 1D data and thus do not refer to image data directly.

The LSTM (MDLSTM) is a multi-dimensional block that only substitutes an LSTM block to RNN block from the MDRNN discussion above. The entries are divided into 3×4 size blocks and are fed into layers of MDSTM, followed by layers of feed-forward ANN. The final output is transformed into a 1D vector and generated using a Connectionist temporal classification (CTC) function [33].

CTC is an algorithm used to perform tasks such as speech recognition and handwriting recognition using only the input and output transcript information. However, no precise alignment information is given; i.e., how a particular audio region for speech or region in handwriting images corresponds to a specific nature. Simple heuristics, such as naming each character, would fail because each character's amount of space varies from person to person and from time to time [34].

10.2.18.1 Demographic and Election Predictions

These words tend to be met by election forecasting. First, elections have direct results that can be measured for the accuracy of the prediction. Such feedback can help forecasters know about wrong prejudices in judgement. Second, theory and empirical evidence about electoral behavior, particularly in the U.S. presidential election, can be used by policy experts to help them read and interpret elections [29].

10.2.18.2 Deep Dreaming

Deep Dream is an experiment that visualizes neural network patterns [38]. Similarly, Deep Dream over interprets and improves patterns in a picture when a child watches clouds and attempts to understand random types.

This is done by transmitting a picture through the network and then measuring the picture's gradient about the layer activations. The image will be changed to increase those activations, improve network patterns, and create a dreamlike image called "Inceptionism." This phase is a reference to the Inception Net, and the movie *Inception* (2010).

10.3 CONCLUSION

This chapter examined the latest state of the art for profound knowledge on the edge of the network. Computer vision, natural language processes, network features, Virtual Reality (VR), Augmented reality (AR), and the need to process end device data in real time were discussed in this chapter. Methodologies were defined for speeding deep learning inferences across terminal devices, edge servers, and the cloud that use the unique DNN models structure and the geospatial location of edge computing user requests.

Significant factors in many works have been found in the trade-off precision, latency, and other performance metrics. Deep learning models were trained, in which multiple end-components worked together on a model of DNN, including techniques to enhance privacy (perhaps with the aid of an edge server or a cloud).

There are still several ongoing challenges in further progress in efficiency, privacy, resource management, benchmarking, and integration with other networking technologies. Technical advances in algorithms, device architecture, and hardware acceleration will solve these challenges.

With the rate of profound learning advancement remaining strong, in addition to current prospects for innovation, it may also present new technological challenges in the complex calculation.

REFERENCES

1. Y. LeCun, Y. Bengio, and G. Hinton, "Deep learning," *Nature*, vol. 521, no. 7553, pp. 436–444, 2015, doi: 10.1038/nature14539.
2. S. Mazumder, "Big Data Tools and Platforms BT - Big Data Concepts, Theories, and Applications," S. Yu and S. Guo, Eds. Cham: Springer International Publishing, 2016, pp. 29–128.
3. G. Muhammad, M. F. Alhamid, M. Alsulaiman, and B. Gupta, "Edge Computing with Cloud for Voice Disorder Assessment and Treatment," *IEEE Commun. Mag.*, vol. 56, no. 4, pp. 60–65, Apr. 2018, doi: 10.1109/MCOM.2018.1700790.
4. L. Huang, S. Bi, and Y.-J. A. Zhang, "Deep Reinforcement Learning for Online Computation Offloading in Wireless Powered Mobile-Edge Computing Networks," *IEEE Trans. Mob. Comput.*, vol. 19, no. 11, pp. 2581–2593, Nov. 2020, doi: 10.1109/TMC.2019.2928811.
5. Z. Lv, D. Chen, R. Lou, and Q. Wang, "Intelligent Edge Computing Based on Machine Learning for Smart City," *Futur. Gener. Comput. Syst.*, vol. 115, pp. 90–99, 2021, doi: https://doi.org/10.1016/j.future.2020.08.037.
6. F. Schuster, B. Engelmann, U. Sponholz, and J. Schmitt, "Human Acceptance Evaluation of AR-Assisted Assembly Scenarios," *J. Manuf. Syst.*, 2021, doi: https://doi.org/10.1016/j.jmsy.2020.12.012.
7. R. Chhabra, S. Verma, and C. R. Krishna, "A survey on driver behavior detection techniques for intelligent transportation systems," in *2017 7th International Conference on Cloud Computing, Data Science Engineering - Confluence*, Jan. 2017, pp. 36–41, doi: 10.1109/CONFLUENCE.2017.7943120.

8. M. P. Véstias, R. P. Duarte, J. T. de Sousa, and H. C. Neto, “Moving Deep Learning to the Edge,” *Algorithms*, vol. 13, no. 5, 2020, doi: 10.3390/a13050125.
9. R. Dong, C. She, W. Hardjawana, Y. Li, and B. Vucetic, “Deep Learning for Hybrid 5G Services in Mobile Edge Computing Systems: Learn From a Digital Twin,” *IEEE Trans. Wirel. Commun.*, vol. 18, no. 10, pp. 4692–4707, Oct. 2019, doi: 10.1109/TWC.2019.2927312.
10. M. McClellan, C. Cervelló-Pastor, and S. Sallent, “Deep Learning at the Mobile Edge: Opportunities for 5G Networks,” *Appl. Sci.*, vol. 10, no. 14, 2020, doi: 10.3390/app10144735.
11. L. Huang, X. Feng, A. Feng, Y. Huang, and L. P. Qian, “Distributed Deep Learning-based Offloading for Mobile Edge Computing Networks,” *Mob. Networks Appl.*, 2018, doi: 10.1007/s11036-018-1177-x.
12. E. Li, L. Zeng, Z. Zhou, and X. Chen, “Edge AI: On-Demand Accelerating Deep Neural Network Inference via Edge Computing,” *IEEE Trans. Wirel. Commun.*, vol. 19, no. 1, pp. 447–457, Jan. 2020, doi: 10.1109/TWC.2019.2946140.
13. Y. He, F. R. Yu, N. Zhao, V. C. Leung, and H. Yin, “Software-Defined Networks with Mobile Edge Computing and Caching for Smart Cities: A Big Data Deep Reinforcement Learning Approach,” *IEEE Commun. Magazine*, vol. 55 (12), pp. 31–37, 2017, doi: 10.1109/MCOM.2017.1700246.
14. M. C. Trivedi and N. K. Trivedi, “Audio Masking for Watermark Embedding Under Time Domain Audio Signals,” in *2014 International Conference on Computational Intelligence and Communication Networks,* 2014, doi: 10.1109/CICN.2014.166.
15. “A Privacy-Preserving Deep Learning Approach for Face Recognition with Edge Computing,” Jul. 2018, [Online]. Available: https://www.usenix.org/conference/hotedge18/presentation/mao.
16. E. Kristiani, C. Yang, and C. Huang, “iSEC: An Optimized Deep Learning Model for Image Classification on Edge Computing,” *IEEE Access*, vol. 8, pp. 27267–27276, 2020, doi: 10.1109/ACCESS.2020.2971566.
17. B. Yang, X. Cao, C. Yuen, & L. Qian, “Offloading Optimization in Edge Computing for Deep-Learning-Enabled Target Tracking by Internet of UAVs,” *IEEE Internet Things J.*, vol. 8 (12), pp. 9878–9893, 2020, doi: 10.1109/JIOT.2020.3016694.
18. P. Subramaniam and M. J. Kaur, “Review of Security in Mobile Edge Computing with Deep Learning,” in *2019 Advances in Science and Engineering Technology International Conferences (ASET)*, Mar. 2019, pp. 1–5, doi: 10.1109/ICASET.2019.8714349.
19. M. Chen, W. Li, Y. Hao, Y. Qian, and I. Humar, “Edge Cognitive Computing Based Smart Healthcare System,” *Futur. Gener. Comput. Syst.*, vol. 86, pp. 403–411, 2018, doi: 10.1016/j.future.2018.03.054.
20. M. Bensalem, J. Dizdarevć, and A. Jukan, “Modeling of Deep Neural Network (DNN) Placement and Inference in Edge Computing,” in *2020 IEEE International Conference on Communications Workshops (ICC Workshops)*, Jun. 2020, pp. 1–6, doi: 10.1109/ICCWorkshops49005.2020.9145449.
21. A. Ndikumana, N. H. Tran, D. H. Kim, K. T. Kim, and C. S. Hong, “Deep Learning Based Caching for Self-Driving Cars in Multi-Access Edge Computing,” *IEEE Trans. Intell. Transp. Syst.*, pp. 1–16, 2020, doi: 10.1109/TITS.2020.2976572.
22. A. Marchisio *et al.*, “Deep Learning for Edge Computing: Current Trends, Cross-Layer Optimizations, and Open Research Challenges,” in *2019 IEEE Computer Society Annual Symposium on VLSI (ISVLSI)*, Jul. 2019, pp. 553–559, doi: 10.1109/ISVLSI.2019.00105.
23. S. R. Sahoo and B. B. Gupta, “Multiple Features Based Approach for Automatic Fake News Detection on Social Networks Using Deep Learning,” *Appl. Soft Comput.*, vol. 100, p. 106983, 2021, doi: https://doi.org/10.1016/j.asoc.2020.106983.

24. H. Khelifi *et al.*, "Bringing Deep Learning at the Edge of Information-Centric Internet of Things," *IEEE Commun. Lett.*, vol. 23, no. 1, pp. 52–55, Jan. 2019, doi: 10.1109/LCOMM.2018.2875978.
25. I. Sharma, R. Tiwari, and A. Anand, "Open Source Big Data Analytics Technique," in *Proceedings of the International Conference on Data Engineering and Communication Technology*, 2017, pp. 593–602.
26. R. Pandey, A. Singh, A. Kashyap, and A. Anand, "Comparative Study on Realtime Data Processing System," in *2019 4th International Conference on Internet of Things: Smart Innovation and Usages (IoT-SIU)*, Apr. 2019, pp. 1–7, doi: 10.1109/IoT-SIU.2019.8777499.
27. A. Luckow *et al.*, "Artificial Intelligence and Deep Learning Applications for Automotive Manufacturing," in *2018 IEEE International Conference on Big Data (Big Data)*, Dec. 2018, pp. 3144–3152, doi: 10.1109/BigData.2018.8622357.
28. S. Ghose and J. J. Prevost, "Enabling an IoT System of Systems through Auto Sound Synthesis in Silent Video with DNN," in *2020 IEEE 15th International Conference of System of Systems Engineering (SoSE)*, Jun. 2020, pp. 563–568, doi: 10.1109/SoSE50414.2020.9130483.
29. Y. Liu, H. Qu, W. Chen, and S. H. Mahmud, "An Efficient Deep Learning Model to Infer User Demographic Information from Ratings," *IEEE Access*, vol. 7, pp. 53125–53135, 2019, doi: 10.1109/ACCESS.2019.2911720.
30. T. M. Ghanim, M. I. Khalil, and H. M. Abba, "Comparative Study on Deep Convolution Neural Networks DCNN-based Offline Arabic Handwriting Recognition," *IEEE Access*, vol. 8, pp. 95465–95482, 2020, doi: 10.1109/ACCESS.2020.2994290.
31. T. Falas and H. Kashani, "Two-Dimensional Bar-Code Decoding with Camera-Equipped Mobile Phones," in *Fifth Annual IEEE International Conference on Pervasive Computing and Communications Workshops (PerComW'07)*, Mar. 2007, pp. 597–600, doi: 10.1109/PERCOMW.2007.119.
32. W. Cho, S.-W. Lee, and J. H. Kim, "Modeling and Recognition of Cursive Words with Hidden Markov Models," *Pattern Recognit.*, vol. 28, no. 12, pp. 1941–1953, 1995, doi: https://doi.org/10.1016/0031-3203(95)00041-0.
33. A. Alahi, K. Goel, V. Ramanathan, A. Robicquet, L. Fei-Fei, and S. Savarese, "Social LSTM: Human Trajectory Prediction in Crowded Spaces," in *2016 IEEE Conference on Computer Vision and Pattern Recognition (CVPR),* Jun. 2016.
34. C. Gao, A. Rios-Navarro, X. Chen, T. Delbruck, and S. Liu, "EdgeDRNN: Enabling Low-latency Recurrent Neural Network Edge Inference," in *2020 2nd IEEE International Conference on Artificial Intelligence Circuits and Systems (AICAS)*, Aug. 2020, pp. 41–45, doi: 10.1109/AICAS48895.2020.9074001.
35. I. Goodfellow, Y. Bengio, and A. Courville, "Deep Learning," MIT Press, 2016.
36. https://www.adobe.com/in/creativecloud/design/discover/colorize-black-and-white-photos.html
37. https://www.infoq.com/news/2020/09/ai-created-foley/
38. https://www.tensorflow.org/tutorials/generative/deepdream

11 Designing an Efficient Network-Based Intrusion Detection System Using an Artificial Bee Colony and ADASYN Oversampling Approach

Manisha Rani
Research Scholar, Department of Computer Science,
Punjabi University
Punjab, India

Gunreet Kaur
Department of Computer Science and Engineering,
Thapar Institute of Engineering and Technology
Punjab, India

Gagandeep
Department of Computer Science, Punjabi University
Punjab, India

CONTENTS

DOI: 10.1201/9781003143468-11

11.1 INTRODUCTION

In the digital world, security has become a critical issue in modern networks over the past two decades. It has led to a rapid increase of security threats along with the growth of computer networks. Despite strong security systems to protect against threats, numerous intruders that can violate the security policies such as Confidentiality, Integrity, and Availability (CIA) of the network still exist. An intruder may attempt to obtain unauthorized access to information, manipulate the information, and make the system unreliable by deploying unnecessary services over the network. Therefore, to secure the network from intruder activities, strong security mechanisms need to be designed (Madhavi 2012). A number of security mechanisms are available, such as cryptographic techniques, firewalls, packet filters, an Intrusion Detection System (IDS), and so on. Subsequently, malicious users use different techniques, such as password guessing, unnecessarily loading the network with irrelevant data, and unloading network traffic, to exploit system vulnerabilities. Therefore, it is highly unrealistic to protect the network completely from breaches. However, it is possible to detect the intrusions so that the damage can be repaired by taking an appropriate action (Kumar and Kukreja 2021). Thus, the IDS plays a significant role in the network security field by providing a solid line of defense against malicious users. It detects any suspicious activity performed by intruders by monitoring network traffic and issues alerts whenever any abnormal behavior is sensed (Rani and Gagandeep 2019). IDS is broadly classified into two types: NIDS and HIDS. NIDS stands for network-based IDS, which analyzes the network traffic by reading individual packets through the network layer and transport layer, whereas HIDS stands for host-based IDS, which monitors every activity of individual hosts or devices on the network. IDS can detect known attacks through signature-based detection and unknown attacks through anomaly-based detection. Both approaches have their own limitations, such as the former detection approach is only good for finding known attacks but not good for unknown attacks because it can match the incoming patterns with the stored patterns only. The system must update the database with a new attack signature whenever a novel attack is identified. Whereas anomaly-based detection can detect both known and unknown attacks, it suffers from high false-alarm rates because of its non-linear nature, high dimensional features, and mixed type of features in the datasets (Kukreja and Dhiman 2020). To overcome these challenges, various machine learning algorithms and evolutionary algorithms exist that are mainly used for classification and feature reduction processes, respectively. Recently, various algorithms, such as random forest, naïve Bayes, k-nearest neighbor, etc., have been used as IDS classifiers. However, the presence of irrelevant features in the dataset weakens the performance of a classifier (Montazeri et al. 2013). Thus, to improve the performance of a classifier, it is very important to reduce the dimensionality of the feature space by identifying and selecting relevant features from the original feature set that are needed for classification. But selecting the relevant features from the full set is itself a challenging task.

Currently, bio-inspired algorithms, such as a genetic algorithm (Sampson 1976), particle swarm optimization (Poli, Kennedy, and Blackwell 2007), ant colony optimization (Dorigo, Birattari, and Stutzle 2006), etc., are emerging techniques used for feature selection because of their high convergence power and searching behavior. They are inspired by the biological nature of animals, insects, and birds and work based upon the principle of their intelligent evolutionary behavior. These algorithms help solve complex problems with improved accuracy and are very useful in finding an optimal solution to a given problem. Although the feature selection approach helps to reduce the false alarm rate of IDS due to the imbalanced nature of IDS datasets, it is unable to improve the performance of classifiers significantly. Generally, IDS datasets contain large amounts of normal data (i.e., majority class instances) when compared to attack data (i.e., minority class instances), which results in a class imbalance problem. It can be addressed either at the data level or classifier level by changing the class distribution of training set itself or by altering the training algorithm rather than training data, respectively. At the data level, various under-sampling techniques, such as one-sided selection and oversampling techniques like Synthetic Minority Oversampling Technique (SMOTE), the clustering approach, etc., are available to balance the instances by inserting some random instances to minority samples or by removing some instances from majority instances, respectively. At the classifier or algorithmic level, two approaches are employed, i.e., either the threshold approach that can be used by adjusting the decision threshold of a classifier or the cost sensitive approach that can be used by modifying the learning rate parameter of algorithm. In this chapter, data is initially preprocessed using the min-max normalization technique. Then, data is balanced using the ADASYN technique (He et al. 2008). Then, a subset of features is chosen from the original set using the ABC algorithm. The optimality of the feature subset is evaluated using the random forest classifier. The performance of IDS is evaluated using the NSL KDD dataset. After an empirical analysis of the classification accuracy of various classifiers, random forest was chosen.

The rest of this chapter is organized as follows: Section 11.2 deals with surveys related to different feature selection and class imbalance approaches. Section 11.3 describes the dataset used in the experimental analysis. Section 11.4 deals with the methodology followed for proposed work. Then experimental results and their comparative analysis are reported in Section 11.5. This chapter concludes with a summary in Section 11.6.

11.2 RELATED WORK

Numerous ML-based classifiers and evolutionary algorithms have been proposed for the classification and feature selection process for IDS. To identify attacks from normal traffic, researchers employ various steps, such as data pre-processing, feature selection and reduction techniques, and classification steps. First of all, existing work related to IDS needs to be reviewed to identify research gaps in previous work and then some new work is proposed for further research.

Tavallaee et al. (2009) analyzed the publicly available KDDCup'99 dataset in order to solve the inherent problems of the dataset. A major problem occurs due to a large number of redundant records; and the level of difficulty in both the training and testing sets leads the ML algorithm to be biased toward majority class instances and there is a large gap between training and testing accuracy. To solve these problems, a new version of KDDCup'99—NSL KDD—is proposed. Although NSL KDD can be used as a benchmark dataset, it still suffers from a few problems, such as data imbalancing, etc. Pervez and Farid (2014) proposed an SVM-based feature selection algorithm and classifier to select a subset of features from the original set. It achieved 82.38% of testing accuracy using 36 features and compared its results with existing work. Ingre and Yadav (2015) evaluated the performance of NSL KDD using a back propagation algorithm of the Artificial Neural Network (ANN) architecture. Before training, data was pre-processed by converting each feature into numerals and normalizing the data into a range [0, 1] using the Z-score approach. Then, 41 features of KDDTrain+ and KDDTest+ were reduced to 23 using the information gain technique and then the reduced features were learned through the neural network. Aghdam, Kabiri, and others (2016) proposed the IDS model by selecting feature subset using the Ant Colony Optimization (ACO) approach followed by the nearest neighbor classifier to identify attacks from normal traffic. Although it achieved better results than existing work, the ACO algorithm suffers from computational memory requirements and low speed due to the separate memory required by each ant. Kaur, Pal, and Singh (2018) compared the performance of the hybridization of K-Means with the Firefly algorithm and Bat algorithm with other clustering techniques and it showed that the proposed work outperformed other techniques with huge margins. However, data preprocessing involving data normalization might also improve the performance of the proposed work. Hajisalem and Babaie (2018) proposed another hybrid model using ABC and Artificial Fish Swarm (AFS) evolutionary algorithms over NSL KDD and UNSW-NB15 datasets. Mazini, Shirazi, and Mahdavi (2019) designed a hybrid model for anomaly-based NIDS using the ABC algorithm and Adaboost algorithm to select features and to classify attacks and normal data using the NSL KDD and ISCXIDS2012 datasets, respectively. Although evolutionary algorithms outperformed various existing works, they could not address the imbalance problem in research work.

Recently, Jiang et al. (2020) designed a hierarchical model based on the deep learning approach by addressing the class imbalance problem using the feature selection process to classify attack data. It combined the one-sided selection (OSS) and SMOTE techniques for handling undersampling and oversampling, respectively. Then, it used Convolutional Neural Network (CNN) to choose spatial features and used Bi-directional long short-term memory (Bi-LSTM) to extract temporal features from the NSL KDD and UNSW-NB15 dataset. It identified attack data from normal by achieving 83.58% and 77.16% testing accuracy using the NSL KDD and UNSW-NB15 datasets, respectively. Alkafagi and Almuttairi (2021) designed Proactive Model for Swarm Optimization (PMSO) for selecting individual optimal feature subsets using Particle Swarm Optimization (PSO) and the Bat algorithm followed by a decision tree classifier. Both studies were successful in improving the performance of the proposed work as compared to existing work but did not address the common issue, i.e., the class imbalance problem. Tao et al. (2021) achieved classification accuracy up to 78.47% by combining a K-means clustering approach with the SMOTE algorithm to equalize the minority data instances with majority instances in NSL KDD dataset, and then using an enhanced RF algorithm for the classification process. Although it addressed the class imbalance, it could not achieve effective results for minority attack data such as R2L and U2R attack types. Priyadarsini (2021) solved the class imbalance problem at the data level using the borderline SMOTE algorithm through the RF technique, and then the selected feature subset using the ABC algorithm. Various classifiers, such as SVM, DT, and K Nearest Neighbor (KNN), were applied for the classification of attack data from normal data using the KDDCup'99 dataset only. Liu and Shi (2022) designed a hybrid model for anomaly-based NIDS using the GA algorithm for feature selection and the RF algorithm for classification. It achieved a high classification accuracy rate using the NSL KDD and UNSW-NB15 training sets but did not test or validate the data results using the testing set of both datasets.

From existing work, it is found that evolutionary algorithms are being extensively used for the feature selection process, followed by ML classifiers. It is also predicted that lesser research has been done to address the class imbalance problem. In this chapter, a data sampling technique, such as ADASYN, has been used for balancing the NSL KDD dataset. Then, the ABC algorithm has been applied for finding the optimal subset of features, followed by the RF classifier to build an effective model for IDS.

11.3 DATASET USED

To empirically analyze the results of the IDS model, widely used dataset, i.e., NSL KDD, has been considered. The experiments are performed using 100% of the dataset. The data is trained over 70% of training set and remaining 30% of the training set is used for validating the results. Then, the results of dataset are tested using the testing set.

11.3.1 NSL KDD Dataset

Due to inherent problems in the KDDCup'99 dataset, the performance of various researchers work was affected. To overcome the issues, an improved version was made and renamed the NSL KDD dataset (Tavallaee et al. 2009). It is the benchmark dataset used by every researcher to compare the performance of IDS models. It contains 42 attributes, of which 41 attributes belong to network flow type, whereas the last 42nd attribute indicates the label assigned to each attribute. All attributes except the last one are categorized into three types: content related, time related, and host-based content. On the other hand, the last attribute is categorized into five classes, where four classes belong to the attack class and one class belongs to the normal data. The four attack classes are Distributed Denial of Service (DDoS), Probe, User to Root (U2R), and Remote-2-Local (R2L). It is publicly available over the internet and consists of the training set and testing set KDDTrain+ and KDDTest+, respectively. A number of instances present in each set are described in Table 11.1:

TABLE 11.1
Record Distribution of NSL KDD Dataset

Dataset	Total Instances	Record Distribution	
		Normal	Attack
KDDTrain+	125973	67343	58630
KDDTest+	22544	9711	12833

11.4 INTRUSION DETECTION PROCESS

The IDS follows some basic steps to detect intrusions from network traffic data. To monitor network traffic, there must be some genuine data containing information about network packets. The first step involves the collection of data from available sources over which the entire detection process has to be done. In this chapter, a benchmark set (i.e., NSL KDD) has been used for the empirical analysis of the data. These datasets contain the metadata of network-related data and different attacks that can be generated by the intruders. The second step involves data pre-processing, which consists of two sub-steps: data conversion and data normalization. Additionally, to address the class imbalance issue, the ADASYN oversampling method is employed to balance the dataset. Then, in order to reduce the dimensions of data, redundant and noisy data is removed from entire dataset and only the relevant features are selected through the feature reduction and feature selection process. Then, the reduced feature set of data is fed into the classifier to discriminate between normal and attack data.

11.4.1 Data Pre-Processing

Because of a vast amount of network packets, the unequal distribution of data in datasets, and the instability of data toward the changing behavior of the network, it is very difficult to classify attack data from normal data with high accuracy rates. Therefore, there is a need to pre-process data before putting it over the detection model. Before classification, the data is pre-processed through data conversion and data normalization.

11.4.1.1 Data Conversion

The NSL KDD contains heterogeneous data, such as numerical and non-numerical types of data. Most of the IDS classifiers accept numerical data types only. So it is necessary to convert all the data into a homogenous form; i.e., a numeric type. First, the data conversion is involved in the pre-processing step, which converts all non-numerical features into integers. For instance, the protocol feature in NSL KDD contains a string type of values, such as TCP, ICMP, and UDP, which are converted into numbers by assigning values, 1, 2, and 3, to them, respectively.

11.4.1.2 Data Normalization

After converting all features into integer types, they can be either discrete or continuous in nature, which are incomparable and may hinder the performance of the model. To improve the detection performance significantly, the normalization technique is applied in order to rescale the dataset values into a range of interval [0, 1]. There are various normalization techniques, such as z-score, Pareto scaling, sigmoidal, min-max, etc., available. Based on the empirical analysis, the min-max normalization approach has been used in this chapter. Every attribute value is rescaled into [0, 1] by transforming the maximum value of that attribute into 1, and the minimum value of that attribute into 0, while the remaining attribute values are transformed into decimal values lying between 0 and 1. For every feature x_i, it is calculated mathematically as follows:

$$X_i = \frac{x_i - min(x_i)}{max(x_i) - min(x_i)} \tag{11.1}$$

where x_i denotes *i*th feature of dataset x, $min(x_i)$ and $max(x_i)$ denote the minimum and maximum values of the *i*th feature of dataset x, respectively. X_i represents the corresponding normalized or rescaled value of the *i*th feature value.

11.4.2 Data Sampling

Generally, large datasets suffer from an unequal distribution of records known as a class imbalance problem. The training set of the IDS datasets contains more normal data, known as majority instances, than attack data, known as minority class instances. For example, the large gap between normal records and attack records is 8713 in KDDTrain+ of NSL KDD, making the results biased toward normal data instances. It is difficult to detect U2R and R2L attacks in NSL KDD because it contains only 52 U2R and 995 R2L instances, which are fewer than DoS and Probe class instances. Therefore, it fails to detect minority attack types in the presence of vast amounts of normal data effectively. In order to give importance to minority class instances, we have used the ADASYN oversampling technique in which the minority instances are oversampled by generating minority data instances based on the density distribution of samples so that it can be equally represented as the majority instances in the dataset (Liu, Gao, and Hu 2021). It uses the KNN approach to add pseudo instances by finding the Euclidean distance between neighbors of minority samples. The pseudo instances are added by joining it with the KNN of each minority sample of the feature vector. The number of neighbors used to introduce pseudo instances in minority instances is five in this chapter. The normalized KDDTrain+ dataset represented by TR_d is given as input to the ADASYN algorithm. It consists of total N number of samples, out of which m_i denotes the number of minority samples and m_a denotes the number of majority samples in the dataset, where $m_i < m_a$.

ALGORITHM 1: ADASYN FOR CLASS IMBALANCE PROBLEM

Input: TR_d: Number of features in set denotes population size
N: Total number of samples
M_i: Number of minority sample
M_a: Number of majority sample
Deg_{Th}: Predefined threshold for maximum tolerated degree of class imbalance
Output: Balanced dataset

Begin

$\boldsymbol{deg} = {}^{\boldsymbol{M_i}}/_{\boldsymbol{M_a}}$ where deg ε (0,1] (11.2)

Check if deg < Deg_{Th}

$\boldsymbol{S} = (\boldsymbol{M_a} - \boldsymbol{M_i}) * \boldsymbol{\beta}$ where $\boldsymbol{\beta}$ ε $[\mathbf{0}, \mathbf{1}]$ (11.3)

$\boldsymbol{R_i} = {}^{\Delta_i}/_{\boldsymbol{K}}\ \forall \boldsymbol{i} \in \boldsymbol{M_i}$ (11.4)

for (each i)

$\boldsymbol{NR_i} = {}^{\boldsymbol{R_i}}/_{\sum_{i=1}^{M_a} \boldsymbol{R_i}}$ (11.5)

i++

end for

for (each i)

$\boldsymbol{S_i} = \boldsymbol{NR_i} * \boldsymbol{S}$ (11.6)

i++

end for

while$(\boldsymbol{S_i})$

$s_i = x_i + (x_{ki} - x_i) * \boldsymbol{rand}(\mathbf{0},\mathbf{1})$

end while

End

This algorithm is mainly based on the density distribution criterion of a normalized ratio such that $\Sigma_i NR_i = 1$. This ratio automatically decides the number of synthetic samples that are needed to be added for each minority sample. The balanced dataset is generated based upon the β coefficient, which defines the desired balance level of the whole dataset after adding synthetic samples in the dataset. The major advantage of using this technique is its capability to generate more synthetic samples for those examples that are difficult to learn. That's why, ADASYN is considered a more efficient approach than SMOTE (Chawla et al. 2002), SMOTEBoost (Chawla et al. 2003), etc. Therefore, a balanced KDDTrain+ set is generated using ADASYN to equally represent the minority and majority class instances.

11.4.3 Feature Selection

Apart from data pre-processing, it is very challenging to monitor large amounts of network traffic that has ambiguous and redundant data. It is very crucial to reduce the dimensions of the entire data in order to secure the network from intrusions in real time. The dimensionality reduction of large datasets further improves the performance of the IDS classifiers of system. It can be achieved by selecting informative or relevant features while rolling out redundant and irrelevant features from the original set through the feature selection process. On the basis of the evaluation criteria, the feature selection is of two kinds: filter-based and wrapper-based feature selection (FS). Filter-based FS assigns weights to features and filters out the irrelvant features based on the order of the weight. Although it saves time finding the method, there is no role for classification algorithm. On the contrary, wrapper-based FS takes into account the effect the classification algorithm has in finding the feature subset, which results in high classification accuracy when compared to filter-based FS (Li et al. 2021). That's why, we prefer the wrapper-based FS approach in this chapter. The basic procedure of wrapper-based feature selection involves selecting the subset of features from the original set and then the generated set is evaluated for its optimality. If the generated set is evaluated to be optimal, it is chosen as the best subset; otherwise, another subset is evaluated. In recent years, swarm intelligence emerged out as an effective approach for feature selection as compared to other approaches. It is inspired by the collective behavior of swarms, such as honey bees, birds, flocks, insects, etc., which interact with each other through specific behavioral patterns. The swarm intelligence-based algorithms are capable of solving non-linear complex problems within less computational time, with low computational complexity (Blum and Li 2008). After studying the pros and cons of various evolutionary techniques, we elected to use the ABC for feature selection in this chapter. ABC has been used to find the optimal subset of features from the NSL KDD dataset.

11.4.3.1 Artificial Bee Colony

ABC is meta-heuristic evolutionary algorithm based on swarm intelligence introduced by Karaboga (2005). It is biologically inspired from the collective behavior of honeybees. It mainly consists of three kinds of bees, employed bees, onlooker bees, and scout bees, that accomplish the task through global convergence. The employed bees search for the food source in the hive on the basis of few properties of food, like the concentration of energy (i.e., nectar amount), closeness to the hive, etc. They share their information about the searched food source with the onlooker bees in the dancing area of the hive. Then the onlooker bees further search for the most profitable food source on the basis of the information acquired from the employed bees in the hive. The location of the most profitable food source is memorized by the bees and they start exploiting that location. On the other hand, the employed bees turn into scout bees when the food source is abandoned. Then, the scout bees start searching for a new food source in a random manner.

In this algorithm, the solution is initiated by a number of employed bees present in the hive. The solution is represented by the number of features contained in the dataset. Initially, the solution in the population denoted by S is defined as $S_i = \{1, 2, 3, \ldots, N\}$, where N denotes the number of attributes and S_i denotes the ith food source. For NSL KDD, the value of N is 41. The values or records contained in the attribute are taken from a balanced set of the dataset. In this chapter, the ABC algorithm has been divided into three phases, as follows:

- **Initialization Phase:** In this phase, the food source S of the population is initialized using following eqn. (11.7):

$$S_{i,j} = l_j + (u_j - l_j) * rand(0,1) \tag{11.7}$$

where l_j and u_j indicate the lower and upper bound of solution $S_{i,j}$, respectively, and $rand(0,1)$ is the random function used to generate the number in the range (0, 1).
- **Employed Phase:** After initializing the food source, the employed bees start searching for another food source in their neighborhood and update the position of food source using eqn. (11.8):

$$NS_{i,j} = S_{i,j} + \emptyset_{i,j} * (S_{k,j} - S_{i,j}) \tag{11.8}$$

where $NS_{i,j}$ denotes the possible candidate food source, $S_{k,j}$ represents the new food source searched by employed bees in the neighborhood of the prior food source $S_{i,j}$, and $\emptyset_{i,j}$ is the random number that lies in the interval range [-1, +1].
- **Onlooker Phase:** After exploring the neighborhood, the employed bees share their information with onlooker bees in the dancing area. Then onlooker bees exploit that food source based on the probability function using eqn. (11.9):

$$Prob_{func(i)} = \frac{fitness_i}{\sum fitness_i} \tag{11.9}$$

where $fitness_i$ denotes the fitness value of ith food source. This fitness value is evaluated by employed bees using eqn. (11.10). Then, the neighborhood is again explored by employed bees using eqn. (11.5). If the new food source is better than the previous food source, it is replaced by new one; otherwise employed bees find another food source for the hive.

$$fitness_i = \begin{cases} \dfrac{1}{O_i + 1}, & O_i \geq 0 \\ 1 + |O_i|, & O_i < 0 \end{cases} \tag{11.10}$$

where O_i represents the objective function of ith food source. When the food source gets abandoned, the employed bee becomes a scout bee and randomly looks for a new food source using eqn. (11.7). In this way, this algorithm is iterated again and again until each and every feature gets explored and the optimal feature subset is generated by the ABC algorithm.

11.4.4 Classification

To classify normal data from attack data and to evaluate the chosen optimal feature subset using the ABC algorithm in the previous steps, the random forest classifier is used in this chapter. It mainly solves the problems based on classification and regression by building multiple decision trees to reduce the noise effect so that more accurate results can be achieved. The multiple decision trees are created randomly using the feature subset generated by the ABC algorithm in the previous steps. Then, it predicts the data as either normal or attack data based upon majority of aggregated votes made from each sub-tree. For binary classification, it usually gives output in 0 and 1 form in such a way that if an attack gets detected, it gives value 1; otherwise, it gives output value 0. The proposed flowchart for the IDS model is shown in Figure 11.1.

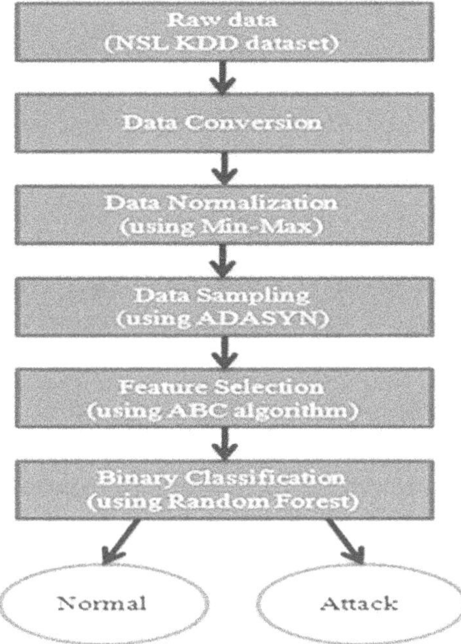

FIGURE 11.1 Flowchart of proposed IDS model.

TABLE 11.2
Parameter Initialization of ABC Algorithm

Parameters	Values
Maximum no. of iterations	100
Population size	10
Population dimension size	N (Feature count) For NSL KDD, N = 42

11.5 RESULTS AND DISCUSSION

The proposed work has been implemented using python language using tensorflow and sklearn, with an Intel Core i5Processor and 8 GB of RAM. For experimental analysis, NSL KDD a widely used dataset has been considered in this chapter. The entire training sets of both datasets were used to train the data, out of which 70% was used for training the algorithm, whereas remaining 30% of the set was used for validating the proposed results. The classification was done using a 10-fold cross validation approach. The features were selected using the ABC algorithm, whose initial values of parameters are defined in Table 11.2.

11.5.1 Performance Parameters

The performance of the proposed work was evaluated based on five parameters, such as classification accuracy, precision, recall, F-score, and receiver operating characteristic (ROC). All these parameters were calculated using the confusion matrix of the model. The confusion matrix was built using the following four indicators:

- True Positive (TP) defines the anomaly records that are correctly recorded as anomaly
- True Negative (TN) defines the normal records that are correctly recorded as normal
- False Positive (FP) defines the normal records that are incorrectly recorded as anomaly
- False Negative (FN) defines the anomaly records that are incorrectly recorded as normal

The performance parameters were calculated using the confusion matrix using the above four indicators as follows:

- **Classification accuracy** is calculated as the ratio of the number of correctly recorded instances to the total number of instances.

$$Accuracy = \frac{TP + TN}{TP + TN + FP + FN} \tag{11.11}$$

- **Precision** is defined as the ratio of the correctly recorded attacks to the total number of instances recorded as attacks.

$$Precision = \frac{TP}{TP + FP} \tag{11.12}$$

- **Recall** is the ratio of the total number of instances that are correctly recorded to the total number of anomaly records in the dataset.

$$Recall = \frac{TP}{TP + FN} \tag{11.13}$$

- **F-score** is the harmonic mean of recall and precision.

$$F - score = \frac{2 * precision * recall}{precision + recall} \tag{11.14}$$

- **ROC** is the receiver operating characteristic curve that records the performance of the binary classification model using the true positive rate (i.e., recall) and false positive rate.

11.5.2 Impact of Class Imbalance

Because of the unequal distribution of class instances in NSL KDD, they suffer from a class imbalance problem. This problem has adverse effects on the performance of the IDS model. Therefore, it is very important to solve this problem using some suitable approaches. It can be solved at either the data level or the classifier level. In this chapter, it is solved at data level using the ADASYN technique. It balances the dataset by oversampling the minority class instances in such a way that both the majority and minority class instances are equally represented in the dataset. The normal and attack data is classified using the random forest classifier without any feature selection process in this section. The classification accuracy without using any imbalance technique is found to be 77.71%, whereas when using the ADASYN technique, it is evaluated to be 78.76%. From the results, it is evident that the accuracy has been improved by 1.05% for NSL KDD, which is of great importance.

11.5.3 Performance of Proposed Work

In order to find the optimal feature subset, we have used the ABC algorithm for feature selection because of its ability for exploration and exploitation. Further, to address the data imbalance problem, oversampling is done using the ADASYN technique. The impact of class imbalance problem at data level is discussed in the previous section. The experimental results are drawn after training the data over 70% of the training set, and then cross-validating it using 10-fold of remaining 30%

TABLE 11.3
The Confusion Matrix Using NSL KDD

(a) Using the Original Set

	Normal	Anomaly
Normal	9424	287
Anomaly	3775	9058

(b) Using the ADASYN and ABC Algorithm

	Normal	Anomaly
Normal	9434	277
Anomaly	4749	8084

validation set after going through 100 epochs. The confusion matrix using the original set without any modification and after applying ADASYN and ABC feature selection for NSL KDD dataset are given in Table 11.3.

It selected 35 attributes out of 42 attributes using the NSL KDD dataset, which shows the dimensionality reduction of the full dataset. With this, it is able to minimize the storage requirements of the program effectively. The list of features selected using the ABC algorithm over NSL KDD is shown in Table 11.4.

The proposed work is evaluated in terms of various parameters, such as classification accuracy, precision, recall, F-score, and ROC curve. The accuracy is tested over the KDDTest+ set of NSL KDD and it is evaluated to be 81.98%, which shows a huge improvement against the original accuracy. Similarly, other parameters have shown tremendous results compared to unbalanced full datasets. Therefore, the data imbalance problem plays a crucial role, along with the feature selection process, in

TABLE 11.4
List of Selected Features Using NSL KDD

Dataset	Number of Features Selected	List of Selected Features
NSL KDD	35	{dur, service, flag, src_bytes, dst_bytes, land, wrong_fragment, urgent, hot, logged_in, num_compromised, root_shell, su_attempted, num_root, num_file_creations, num_shells, num_access_files, num_outbound_cmds, is_hot_login, is_guest_login, count,srv_serror_rate, rerror_rate, srv_rerror_rate, same_srv_rate, diff_srv_rate, srv_diff_host_rate, dst_host_count, dst_host_same_srv_rate, dst_host_diff_srv_rate, dst_host_srv_diff_host_rate, dst_host_serror_rate, dst_host_srv_serror_rate, dst_host_rerror_rate, dst_host_srv_rerror_rate}

TABLE 11.5
Proposed Results for Binary Classification

	NSL KDD	
Parameters (in %)	**Original Dataset**	**ADASYN+ABC**
Accuracy	77.70	81.98
Precision	96.68	96.92
Recall	62.99	70.58
F-score	76.28	81.68
ROC	96.48	97.00

classifying normal and attack data in the IDS model. The ROC curve values are also above 95% over the dataset, which shows the proposed work is superior. The empirical results for the NSL KDD dataset are shown in Table 11.5.

The performance of the proposed approach is also compared with existing literature in Table 11.6. From the comparison table, it is evident that the proposed work has shown great improvement compared to previous work.

11.6 CONCLUSION

In this chapter, an effective NIDS is designed using the swarm intelligence-based ABC optimization algorithm followed by a random forest classifier for binary classification. Due to the large size of IDS datasets, they usually suffer from a class imbalance problem, which becomes a crucial barrier to the system performance. To address this issue, data is balanced using the ADASYN oversampling technique in order to give importance to minority instances in the dataset. The impact of class imbalance is effectively evaluated based upon an empirical analysis of classification accuracy in this chapter. In order to improve the performance of the system to a large extent, data is pre-processed using the min-max normalization technique followed by data balancing using the oversampling approach. Later on, the dimensions of large size datasets are reduced by extracting relevant feature set using the ABC optimization

TABLE 11.6
Comparison of Proposed Work with Existing Literature

References	**Classification Accuracy**
Ibrahim, Basheer, and Mahmod (2013)	75.49%
Li et al. (2017) (using GoogLeNet)	77.04%
Li et al. (2017) (using ResNet 50)	79.14%
Al-Yaseen (2019)	78.89%
Tao et al. (2021)	78.47%
Rani and Gagandeep (2021)	80.83%
Proposed work	81.98%

algorithm followed by the random forest classifier. The classification accuracy is calculated to be 81.98% through empirical analysis using the NSL KDD dataset. The proposed work has successfully outperformed the existing literature. Future work will focus on reducing the computational time of the ABC algorithm as it takes long time to train the large datasets, which is very time-consuming.

REFERENCES

Aghdam, Mehdi Hosseinzadeh, Peyman Kabiri, and others. 2016. "Feature Selection for Intrusion Detection System Using Ant Colony Optimization." *International Journal of Network Security*, 18 (3): 420–32.

Al-Yaseen, Wathiq Laftah. 2019. "Improving Intrusion Detection System by Developing Feature Selection Model Based on Firefly Algorithm and Support Vector Machine." *IAENG International Journal of Computer Science*, 46 (4).

Alkafagi, Salam Saad, and Rafah MAlmuttairi. 2021. "A Proactive Model for Optimizing Swarm Search Algorithms for Intrusion Detection System." *Journal of Physics: Conference Series*, 1818:12053.

Blum, Christian, and Xiaodong Li. 2008. "Swarm Intelligence in Optimization." In *Swarm Intelligence*, 43–85. Natural Computing Series (NCS) Springer.

Chawla, Nitesh V, Kevin W Bowyer, Lawrence O Hall, and W Philip Kegel Meyer. 2002. "SMOTE: Synthetic Minority Over-Sampling Technique." *Journal of Artificial Intelligence Research*, 16: 321–57.

Chawla, Nitesh V, Aleksandar Lazarevic, Lawrence O Hall, and Kevin W Bowyer. 2003. "SMOTEBoost: Improving Prediction of the Minority Class in Boosting." In *European Conference on Principles of Data Mining and Knowledge Discovery*, 107–19.

Dorigo, Marco, Mauro Birattari, and Thomas Stutzle. 2006. "Ant Colony Optimization." *IEEE Computational Intelligence Magazine*, 1 (4): 28–39.

Hajisalem, Vajiheh, and Shahram Babaie. 2018. "A Hybrid Intrusion Detection System Based on ABC-AFS Algorithm for Misuse and Anomaly Detection." *Computer Networks*, 136: 37–50.

He, Haibo, Yang Bai, Edwardo A Garcia, and Shutao Li. 2008. "ADASYN: Adaptive Synthetic Sampling Approach for Imbalanced Learning." In *2008 IEEE International Joint Conference on Neural Networks (IEEE World Congress on Computational Intelligence)*, 1322–28.

Ibrahim, Laheeb M, Dujan T Basheer, and Mahmod S Mahmod. 2013. "A Comparison Study for Intrusion Database (Kdd99, Nsl-Kdd) Based on Self Organization Map (SOM) Artificial Neural Network." *Journal of Engineering Science and Technology*, 8 (1): 107–19.

Ingre, Bhupendra, and Anamika Yadav. 2015. "Performance Analysis of NSL-KDD Dataset Using ANN." In *2015 International Conference on Signal Processing and Communication Engineering Systems*, 92–96.

Jiang, Kaiyuan, Wenya Wang, Aili Wang, and Haibin Wu. 2020. "Network Intrusion Detection Combined Hybrid Sampling with Deep Hierarchical Network." *IEEE Access*, 8: 32464–76.

Karaboga, Dervis. 2005. "An Idea Based on Honey Bee Swarm for Numerical Optimization."

Kaur, Arvinder, Saibal K Pal, and AmritPal Singh. 2018. "Hybridization of K-Means and Firefly Algorithm for Intrusion Detection System." *International Journal of System Assurance Engineering and Management*, 9 (4): 901–10.

Kukreja, Vinay, and Poonam Dhiman. 2020. "A Deep Neural Network Based Disease Detection Scheme for Citrus Fruits." In *2020 International Conference on Smart Electronics and Communication (ICOSEC)*, 97–101.

Kumar, Deepak, and Vinay Kukreja. 2021. "N-CNN Based Transfer Learning Method for Classification of Powdery Mildew Wheat Disease." In *2021 International Conference on Emerging Smart Computing and Informatics (ESCI)*, 707–10.

Li, Xin, Peng Yi, Wei, Yiming Jiang, and Le Tian. 2021. "LNNLS-KH: A Feature Selection Method for Network Intrusion Detection." *Security and Communication Networks*, 2021. Hindawi.

Li, Zhipeng, Zheng Qin, Kai Huang, Xiao Yang, and Shuxiong Ye. 2017. "Intrusion Detection Using Convolutional Neural Networks for Representation Learning." In *International Conference on Neural Information Processing*, 858–66.

Liu, Jingmei, Yuanbo Gao, and Fengjie Hu. 2021. "A Fast Network Intrusion Detection System Using Adaptive Synthetic Oversampling and LightGBM." *Computers & Security*. 102289.

Liu, Zhiqiang, and Yucheng Shi. 2022. "A Hybrid IDS Using GA-Based Feature Selection Method and Random Forest." *International Journal of Machine Learning and Computing*, 12 (2).

Madhavi, M. 2012. "An Approach for Intrusion Detection System in Cloud Computing." *International Journal of Computer Science and Information Technologies*, 3 (5): 5219–22.

Mazini, Mehrnaz, Babak Shirazi, and Iraj Mahdavi. 2019. "Anomaly Network-Based Intrusion Detection System Using a Reliable Hybrid Artificial Bee Colony and AdaBoost Algorithms." *Journal of King Saud University-Computer and Information Sciences*, 31 (4): 541–53.

Montazeri, Mohadeseh, Mitra Montazeri, Hamid Reza Naji, and Ahmad Faraahi. 2013. "A Novel Memetic Feature Selection Algorithm." In *the 5th Conference on Information and Knowledge Technology*, 295–300.

Pervez, Muhammad Shakil, and Dewan Md Farid. 2014. "Feature Selection and Intrusion Classification in NSL-KDD Cup 99 Dataset Employing SVMs." In *the 8th International Conference on Software, Knowledge, Information Management and Applications (SKIMA 2014)*, 1–6.

Poli, Riccardo, James Kennedy, and Tim Blackwell. 2007. "Particle Swarm Optimization." *Swarm Intelligence* (1): 33–57.

Priyadarsini, Pullagura Indira. 2021. "ABC-BSRF: Artificial Bee Colony and Borderline-SMOTE RF Algorithm for Intrusion Detection System on Data Imbalanced Problem." In *Proceedings of International Conference on Computational Intelligence and Data Engineering: ICCIDE 2020*, 15–29.

Rani, Manisha, and Gagandeep. 2019. "A Review of Intrusion Detection System in Cloud Computing." In *Proceedings of International Conference on Sustainable Computing in Science, Technology and Management (SUSCOM), Amity University Rajasthan, Jaipur-India*.

Sampson, Jeffrey R. 1976. "Adaptation in Natural and Artificial Systems (John H. Holland)." *Society for Industrial and Applied Mathematics SIAM Review*, Bradford Books, 18 (3): 529–30.

Tao, Wu, Fan Honghui, Zhu HongJin, You CongZhe, Zhou HongYan, and Huang XianZhen. 2021. "Intrusion Detection System Combined Enhanced Random Forest with SMOTE Algorithm." In *2013 International Conference on Cloud & Ubiquitous Computing & Emerging Technologies*, Pune, India.

Tavallaee, Mahbod, Ebrahim Bagheri, Wei Lu, and Ali A Ghorbani. 2009. "A Detailed Analysis of the KDD CUP 99 Data Set." In *2009 IEEE Symposium on Computational Intelligence for Security and Defense Applications*, 1–6.

Index

F

G

H

I

K

L

M

N

O

P

Milton Keynes UK
Ingram Content Group UK Ltd.
UKHW031531071024
449327UK00005B/133